NICK BAKER'S

BUG BOOK

Discover the World of the Mini-Beast!

BLOOMSBURY

LONDON · NEW DELHI · NEW YORK · SYDNEY

The Wildlife Trusts

The Wildlife Trusts are the UK's largest people-powered organisation caring for all nature – rivers, bogs, meadows, forests, seas and much more. We are 47 Wildlife Trusts covering the whole of the UK with a shared mission to restore nature everywhere we can and to inspire people to value and take action for nature for future generations.

Together we care for thousands of wild places that are great for both people and wildlife. These include more than 760 woodlands, 500 grasslands and even 11 gardens. On average you're never more than 17 miles away from your nearest Wildlife Trust nature reserve, and most people have one within 3 miles of their home. To find your nearest reserve, visit **wildlifetrusts.org/ reserves**, or download our free Nature Finder iPhone app from the iTunes store. You can also find out about the thousands of events and activities taking place across the UK – from bug hunts and wildplay clubs to guided walks and identification courses – on the app or at **wildlifetrusts.org/whats-on**

We work to connect children with nature through our inspiring education programmes and protect wild places where children can spend long days of discovery. We want children to go home with leaves in their hair, mud on their hands and a little bit of nature in their heart. Find out more about our junior membership branch Wildlife Watch and the activities, family events and kid's clubs you can get involved with at **wildlifewatch.org.uk**

Our goal is nature's recovery – on land and at sea. To achieve this we rely on the vital support of our 800,000 members, 40,000 volunteers, donors, corporate supporters and funders. To find the Wildlife Trust that means most to you and lend your support, visit **wildlifetrusts.org/ your-local-trust**

The Wildlife Trusts
The Kiln, Mather Road, Newark, Nottinghamshire NG24 1WT
t: *01636 677711*
e: *info@wildlifetrusts.org*

Registered Charity No 207238

wildlifetrusts.org

Find us on
Twitter – **@wildlifetrusts**
Facebook – **facebook.com/wildlifetrusts**

NICK BAKER'S

BUG BOOK

First published in 2006 by New Holland Publishers (UK) Ltd
This edition published 2015 by Bloomsbury Publishing Plc,
50 Bedford Square, London WC1B 3DP

Bloomsbury Publishing Plc
50 Bedford Square
London
WC1B 3DP

www.bloomsbury.com

BLOOMSBURY and the Diana logo are trademarks of Bloomsbury Publishing Plc

A CIP catalogue record for this book is available from the British Library

ISBN (paperback) 978-1-4729-1379-1

Printed in China by C&C Offset Printing Co., Ltd.

10 9 8 7 6 5 4 3 2 1

FSC
www.fsc.org

MIX
Paper from
responsible sources
FSC® C008047

Front cover: Shutterstock
Back cover: Tony Cobley Photography, www.tonycobley.com

Contents

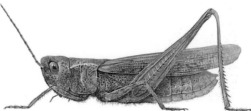

Introduction

If you're irritating your family, annoying the cat or just plain bored, then I've got some therapy for you. All around us are marvellous mini-beasts that are completely overlooked. Get down and dirty your knees and those dots with legs and wings become tottering robots and oozing aliens. It's not science fiction, it's science fact! And it can all be seen in the average park or garden.

These creatures fight battles with amazing weapons, nasty chemicals and terrific tactics. Their weird and colourful world is full of fascinating mechanisms, stunning shapes and groovy gadgets.

A world of tiny animals is dancing in the wind, mooching in leaf litter and playing in ponds. These guys may be small, but their effects (and not just the bad ones) are noticed by us all.

Who rules the world? They do!

Like 'em or not, these wee beasties run our planet. Without their composting services we would be knee deep in dead leaves, dung and dead animals. Nice thought!

Beware of the 'bug effect'

You will often find that if you show someone an insect or spider they will run off, waving their arms about and making more noise than the size of the creature seems to justify. Sadly this easy trick does nothing for the reputation of the harmless creature.

Don't fall into the temptation of scaring the living daylights out of people. Instead try to understand that not everyone likes invertebrates and some people may not understand your love of them. You can try to talk people around – I even managed to get my mum, who is seriously scared of spiders, to hold one of my pet tarantulas.

Left: With the right equipment and a little bit of patience you can zoom in on the world of bugs.

This book will help you realise the place that mini-beasts have in the whole scheme of things. Even if you're not a big fan of worms you can be sure that you love something that eats them, whether it's a badger or a bird.

Remember that it's the little things that make the world go around and think twice about squashing that beetle with your boot as it scampers across your path!

Above: Hoverflies in a poppy. Just peer into any flower and you might find a pollination party a blossoming romance, or even the occasional murder!

About this book

Bug, mini-beast and creepy-crawly are all names used to describe an enormous number of smallish life forms that scuttle, ooze, creep, fly and crawl about the planet. These names can refer to a huge range of different animals found on land or in fresh water, from true six-legged insects to centipedes, spiders and even slugs and snails.

Between them, they look like a collection of extras from a science fiction movie. They range in size from the microscopic nematode worms to the 17 cm long giant, the Goliath Beetle. Some have one foot, others have six, eight, fourteen or even seven hundred and fifty legs! Some have bodies made of segments and others are encased in an armoured shell.

The one thing they all have in common is that none of them has a single bone in their bodies. Scientists call them 'invertebrates', which means 'without a back bone'.

To write about every invertebrate you might stumble across I would have to write for many lifetimes without ever once stepping outside. Where's the fun in that? Invertebrates are out there to be enjoyed. That is, I hope, what this book is about: having fun with bugs.

There are tips about recognising groups of animals and telling them apart from others but that is about as complicated as I'm going to get. I've concentrated on little tricks that I have collected during my life as a 'bug man', which I hope will help you make sense of their sometimes complicated little lives.

Azure Damselfly

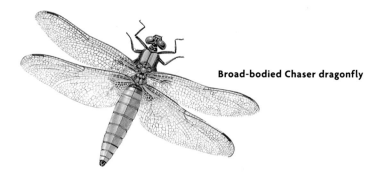

Broad-bodied Chaser dragonfly

The first section of the book is about animals that have no legs, such as worms, slugs and snails.

The next section contains animals called arthropods (which have jointed legs but are not insects), such as woodlice, centipedes, millipedes, spiders and their relatives.

The final part is about insects, including butterflies, moths, dragonflies, damselflies, crickets, grasshoppers, wasps, bees, ants, bugs, beetles and flies.

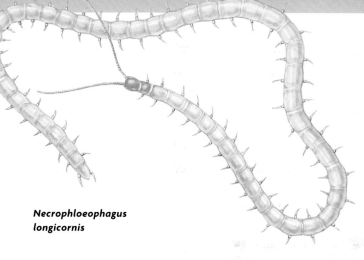

Necrophloeophagus longicornis

The vast variety of mini-beasts you could find in your garden

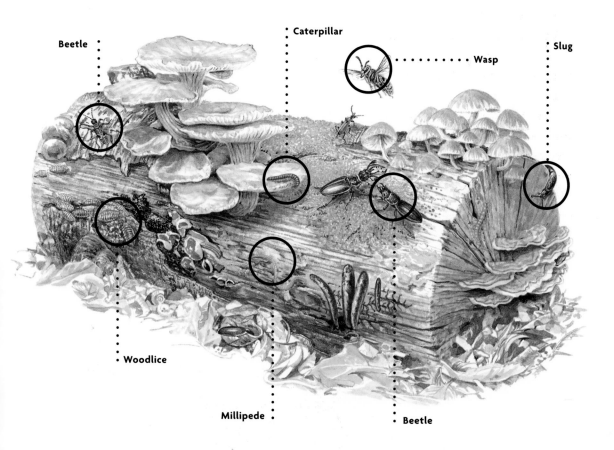

Beetle

Caterpillar

Wasp

Slug

Woodlice

Millipede

Beetle

The Essential Bug-hunting Kit

One of the best things about invertebrates is that they are everywhere. If you are into bugs you will never, ever be bored again, I guarantee it!

Whether you're in a car park or a nature reserve, bugs will be there to marvel at and study. And a great thing about bugs is that they are cheap to get to know. You really don't need any expensive equipment. Just make sure you always carry a notebook and pencil to make notes and sketches.

A few rules ...

Insects, arachnids and their kin get a rough deal from us humans. Some people will kill them without so much as a second thought, but as a budding bug scientist you must make sure that you treat invertebrates with respect.

Be gentle with them, let them go once you

KEEPING YOUR BUG HAPPY

1 Be calm and quiet. You will see more, and small animals may not leg it for the nearest bush, hole or tree!

2 Always return stones and logs to their original position. How would you like it if a giant got you out of bed early, picked you up and then left your house on its side or in the next stree? Also if you are returning an animla to its hidden home put the stone, rock or log in place first and place your mini-beast next to it - it will find its own way back and it won't get accidentally squashed.

3 If you set a trap, particularly a pitfall trap (see page 114), make sure you check it at least once a day to save any bugs you trap from dying.

4 If you hold on to a min-beast for study, take care of it – make sure there is moisture, suitable food, shelter and something absorbent to soak up water droplets or condensation (many bugs die by drowing).

5 Never leave any captive mini-beast – even those that seem to like the heat – in direct sun. Try and avoid any extremes in temperature.

6 Invertebrates can live on very little oxygeb and will survive in air-tight observation pots for quite a long time. However, if you are keeping them for any more than an hour, make sure there is plenty of ventilation. If you are keeping them for long-term study, remember it's better to look a few well than many badly!

7 Avoid holding or touching invertebrates with your hands – damage can be caused by clumsy fingers. Scales and hairs rub off and the animal may die if this happens.

8 Keep a record of what you've found by taking notes and making sketches of your mini-beast.

9 Always return animals to exactly where you found them.

have finished observing their lives and disturb them as little as possible. Follow the guidelines below to keep you and your bugs happy.

Your eyeball ... the best, most valuable, irreplaceable bug-hunting tool

The most important tool for the bug hunter is not his or her expensive butterfly net, nor the glossiest field guide, nor even a super-duper hand lens ... oh no. The single most useful tool is staring right back at you when you look in the mirror. That's right – to be a good mini-beast scientist you need, above all else, to be able to use your eyes!

This may seem easy, but you would be surprised how many people can go for a walk in a meadow and see nothing, while the well-trained observer can spend all day staring at a square metre of lawn and never be bored once! Never underestimate the power of your eyes.

Up close and personal

To appreciate how furry a spider is or watch a woodlouse breathe you need a magnifying hand lens. Of all the gear you could use, this is the only thing you cannot make and it is worth buying a good one.

GETTING BUGS IN PERSPECTIVE

Try lying in the long grass of a hay meadow on a sunny summer or spring day and looking through the grass blades. At first you will see nothing but grass, but slowly you will feel a change in your focus. Instead of a few grass stems you will start to see a wild jungle of plants – ladybirds murdering greenflies, odd-looking bugs with strange mouthparts prodding and poking, giant stilt-walking spiders and harvestmen spanning gaps in the canopy while ants busily shuttle seeds and insects up and down from the dark 'forest floor' to the top of the stems – the place is alive. Stand up and think about it, think of that little patch and then look out in front of you – there is an awful lot more grass out there and many more strange life forms to discover. You are now using your eyes correctly and are ready to embark on your exploration of the creepy crawly world!

There are many different kinds: high magnification (x20), low magnification (x5), big 'Sherlock Holmes' types, small pocket ones, metal or plastic, some with glass lenses, some with plastic and some with a choice of magnification.

I prefer the small metal ones that fold up, especially those with two lenses. You can put them on a string around your neck so that they're always to hand.

Left: Looking at bugs is all about how you see the world. All sorts gadgets and tricks can help.

Making a bug restrainer

YOU WILL NEED:

2 plastic drinking cups • polystyrene tile • marker pen • scissors • PVA glue • cling-film • elastic bands/tape • hand lens

1: Take one of the cups and cut the middle out of the bottom leaving a rim. This is cup 'A'. Stretch a piece of cling-film over the base and the hole and secure it with either tape or elastic bands.

2: Take the other cup, cup 'B', and draw around its base onto the polystyrene tile, then using the scissors cut out this disc.

3: Using the PVA glue stick the polystyrene disc to the underside of the base of cup B.

4: Once the glue is dry, your bug restrainer is complete and ready to use. Simply place your hyperactive bug into cup A and then slowly and gently push cup B into cup A until the bug is sandwiched between the cling-film and the polystyrene. You can now investigate your specimen. It may take a couple of attempts to get your specimen in the best position.

A FEW USEFUL TIPS

• Once the bug is restrained, work quickly. Release the bug as soon as possible because cling-film can suffocate small animals.

• Always make sure your bug is dry, otherwise it will stick and get in a right mess. You may also get condensation, which will spoil your view.

• Never try to restrain a bug that looks like it will be bigger than the base of the cups when stretched out.

The kit ...

Pin 'em down!

As you begin studying mini-beasts you will soon realise that many of them move fast! Sometimes even if you have the creature captive in a pot, it will sprint about so quickly it's hard to tell what colour it is, let alone how many legs it's got.

What you need is a bug restrainer. I know it sounds like a torture device, but it is quite harmless and gentle if used correctly. It can help both you and the bug – you get to look at it close up and the animal gets released quicker because you are not going to hold it for hours in a jam jar.

A couple of points

Tweezers or forceps are useful things. For the bug hunter they are best for gently moving things – turning over a leaf to look on the other side, picking an empty spider skin out of a web, or lifting an animal such as a snail, beetle or caddisfly nymph.

Tweezers are not good for handling small or delicate animals, because it is hard to judge just how much force you are using. For simply touching, unfolding and gently poking, carefully use cocktail sticks or large sewing needles.

Nets please!

Nets are like an extension of your body, allowing you to catch things when you wouldn't stand a chance using your hands and arms.

Watch and learn about your subjects first and use a net only if you really have to, for example to identify something that never seems to settle down, to reach up high, or to collect creatures that live in thick grass or even under water.

Pots of pots

Bug hunting and pots go together like sharks and teeth. Start collecting them now – you will never have enough. Clear straight-sided boxes are always the best for observation, but anything will do: clear takeaway boxes, disposable milkshake cups from the burger joint and margarine containers.

Realising the potential of packaging will become one of your well honed skills as a bug enthusiast. Have fun, experiment, improvise and enjoy recycling!

DIFFERENT NETS FOR DIFFERENT JOBS

- A strong white net with a heavy frame makes a good sweep net, for 'swooshing' through grass or bushes.
- Lightweight black mesh makes a good butterfly or flying insect catching net.
- A thick strong net with big holes for drainage is a good pond net.
- A net with a jam jar placed at the bottom is an adaptation for collecting tiny pond animals and plankton.
- You can buy all of these from specialist shops but, with a little effort, a sewing kit and a garden cane, you should be able to come up with something. Be prepared to improvise: sieves and tea strainers make good pond nets as well as kitchen utensils!

Making a pooter

YOU WILL NEED:

small jar with plastic lid (the kind you get herbs in is ideal) • 50 cm of clear plastic tubing – 5 mm in diameter (from wine-making/DIY shops) •insulating tape • elastic band • a small piece of muslin or fine meshed fabric • scissors • candle • matches • meat skewer

1: Take the pot and in a well ventilated place heat the tip of the skewer with the candle flame. This is a little tricky so ask an adult to help you if you need to. Use the hot tip of the skewer burn two holes, slightly less than the diameter of your plastic tubing (the tubes must fit snugly into the holes).

2: Cut the plastic tubing into two pieces, one 20 cm long, the other 30 cm long. Insert each into one of the holes in the plastic lid (run them under a hot tap if you have difficulties, it makes the plastic softer). Push them in so that they hang about half way into the jar.

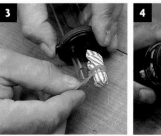

3: On the shorter tube, tie a piece of muslin over the end in the jam jar. Secure it with an elastic band.

4: Put the lid back on the jar. Wrap a small band of coloured tape around the end of the short bit of tubing – this is to remind you which tube to suck on. If you suck on the other one by mistake you could end up with a mouth full of bugs!

5: Test your pooter by picking up small pieces of paper. Suck on the end of the short tube, while directing the end of the long tube over the paper. Working a bit like a vacuum cleaner, a short, sharp, gentle suck will transport the paper into the jar, the fabric on the end of the tube stops it going in your mouth.

One for the little suckers

The trouble with creepy-crawlies is that a lot of them are minuscule and so they are tricky to pick up and look at. Even careful use of tweezers can cause damaged or missing limbs and wings.

Another problem is that, like buses, you often get lots of them at the same time and then they are more than likely to be running in all directions. In situations like these you need a serious bit of kit with a funny name: a pooter.

You can buy them but they can also be easily made for pennies. Once you have made one of these, you can use it again and again.

Spoon and paint brush

Many bugs have a habit of folding up their legs and plummeting to the ground at the slightest disturbance A cool trick here is to use a table spoon to catch the evasive bug. If you also use a small paint brush to sweep delicate creatures off plants or the ground then you are bound to catch something.

If you're squeamish about picking up slugs and snails you can use two spoons to pick them up.

A GUIDE TO GOOD 'POOTERING'

● Only suck up small animals in your pooter. Anything fragile, with long legs or with a body nearly as wide as the tube will get stuck, be damaged or die.

● Try to keep the collection jar dry. Do not blow into the tubes as this fills it up with moisture from your breath.

● Wet and slimy animals such as slugs and snails are not good to 'poot' up as their mucus makes everything sticky.

● Try to keep the number of animals in the collecting jar to a minimum. Keep predators such as spiders away from other animals, as they could well eat them.

● Suck gently. Soft, fragile animals can be injured if they are sucked up too violently – treat them gently and use short, gentle sucks to get the pooter to work best.

Worms

You have probably seen loads of worms and not so much as blinked an eyelid at one, let alone ever been amazed by it. But despite their small size and slimy appearance worms are very, very successful animals. Worms have been making burrows and holes in planet Earth for some 120 million years!

MILES OF BURROWS

Every acre of grassland can contain up to three million earthworms and each year between them they will turn over and mix up ten tons of soil! Every day this number of worms could make nine miles of burrows!

Celebrity status

These lean tunnelling machines are vital to the health of your garden and other ecosystems on the planet. This important attribute of the earthworm has been shouted about by many famous fans.

Just some of the big names a worm could drop include Charles Darwin, who wrote a book about them, the Greek philosopher, Aristotle, who called worms the 'intestines of the soil' and even the Egyptian queen, Cleopatra, who thought worms were sacred (my kind of girl!).

The reason for this high-powered following is that all these people realised that worms play an essential role in keeping the very soil we stand on in good condition. When the soil is good so is the plant life and everything that feeds on it.

As well as improving the quality of soil by breaking down plant material and allowing air and water to pass more freely through it, worms provide food for many furry and feathery animals.

Next time you see a worm struggling across a path, just remember that the animal you are looking at is not just a worm but a living, breathing, rotovating, composting, fertilising, soil shifting, drainage engineer!

Left: Brandling worms. As well as the recycling, drainage and soil engineering work they do, they are dinner for many exciting birds and mammals.

Right: I've just made a wormery – soon I'll have a private window into their secret underworld.

What is a worm?

Earthworms belong to a group of animals known as annelids, the segmented worms. As their name suggests, all of these animals have a body divided into segments. This group includes animals you might find at the seaside, such as ragworms, and also the pond-loving leeches.

The segments on an earthworm can be seen very easily, dividing the body so it looks a bit like a vacuum cleaner tube. Being long, and flexible, worms can squeeze between particles of soil or vegetation. They can also move forward by a series of muscular bulges.

The body appears slimy because it is covered in mucous. In the same way as oil allows a piston to slide easily up and down in its case, mucous helps the worm move through its burrows. It also binds together loose soil particles and it slows down the loss of water through the skin when conditions are dry.

Hairy worms!

Have you ever seen a blackbird, thrush or robin struggling to pull a worm from its burrow in the lawn? How can the worm put up such a fight?

Next time you find an earthworm, try holding it in your gently clenched fist – you will feel just how strong it is as it tries to force its way between your fingers. It uses all those muscles to push and pull its body through the soil. But that isn't the whole story...

Left: Earthworms have a few survival tricks to cope with bigger creatures trying to pull them out of their burrows.

The worm's body

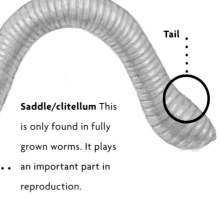

Head or tail? Because worms do not have a 'face' as such, they have no eyes or ears and no obvious nose. The easiest way to tell a worm's head from its tail is to watch which way it crawls. If your worm is inactive then the thin round end is the head and the flattened paddle-shaped end is the tail.

Tail

Saddle/clitellum This is only found in fully grown worms. It plays an important part in reproduction.

Mouth The only recognisable feature on a worm's 'head', and you have to look very closely to see it, the mouth lies just under the tip of its snout.

Sensitive souls You can't see them but a worm has lots of tiny 'taste buds' all over its body – particularly at the head end – something like 700 per square millimetre. That is why worms are able to find food, despite not having any eyes. They are also very sensitive to touch and have many nerve endings just under their skin. Their bristles double up as vibration sensors. Worms also have light-sensitive cells on their bodies.

CHARMING WORMS

Worm-eating animals have lots of different ways of catching these subterranean beasts, but one of the most ingenious is one you can try for yourself.

'Puddling' is behaviour that certain smart birds, particularly gulls, engage in. Watch them on damp short grass, on overcast days – playing fields are a good place to keep an eye open for this behaviour. The birds appear to be doing an on-the-spot jig, stepping up and down on the same spot repeatedly. What they are actually up to is convincing the worms to come to the surface. The stamping produces vibrations to which the worms are very sensitive. To the worm this 'drumming' could sound a bit like rain.

Worms tend to surface in the rain perhaps because conditions above the surface suddenly become better for them or simply to escape water-logged burrows. Whatever the reason, the come to the surface and become dinner.

Bizarrely, humans also 'puddle' – not for food but for a competition known as 'worm charming'. This involves stamping, patting, prodding and even watering the soil to get as many worms to surface as possible. Try it on your lawn or in the park – you may have to wait a while and people may ask questions but eventually you should be rewarded by a worm or two poking its head out of the grass!

Earthworms are grouped together with some 3,000 other worms known as *Oligochaetes*, which means 'few bristles'. If you place a worm on a piece of paper, you may be surprised at the noise it makes as it moves; not a squishy slurpy noise but a dry scratchy one. If you gently pull a worm backwards between your fingers it will feel rough.

Both of these things are caused by lots of tiny bristles (four per segment) called setae, which help the worm move through the soil. It is these little bristles that give worms a fighting chance in that deadly tug-o-war with garden birds!

Tunnel wizards

Place a worm on the soil and within minutes it has forced its way below, into its dark damp world. Despite not having any arms and legs,

Above: Not shy, earthworms will do it on your lawn in full view of all that tread quietly and are armed with a red torch.

worms progress through the soil in a number of ways.

Their body is very flexible and stretchy. It is also very muscley. In fact your average *Lumbricus terrestris* (Common Earthworm), probably has more muscles than Arnold Schwarzenegger!

Lots of ring-shaped muscles run around the body and long ones run along its length. It can control these muscles in a very precise way. Each segment is like a tough balloon full of water. When you push in one place the balloon expands and squishes out another way.

The worm can make itself long and thin,

squeeze into tight gaps between soil particles, and then use its muscles and that 'water balloon' effect to squash its segments up, causing its body to push apart the soil. By repeating this action, and using the bristles to grip the soil, the worm can push and pull its way underground, and it smoothes its path by producing mucous.

X-rated annelid

How can you tell whether your earthworm is a he or a she? It's very easy indeed. All earthworms are both! They are hermaphrodite, having both male and female reproductive parts. Despite this, one worm does need another.

Have a look at your worms – do they have what looks like a big, fat, often pale segment approximately two thirds of the way down the body? If so they are grown ups. Only mature worms have a 'saddle' like this, also called a clitellum. It's used as part of the worms' slimy embrace, when they meet and mate, producing a blanket of thick mucous. If you go out on warm, wet summer nights armed with a red cellophane-covered torch, search the lawn for worms lying next to each other, (it will usually be the big Common Earthworm *Lumbricus terrestris*, as many other species mate underground). Tread carefully as you go about your business as they are really, really sensitive to vibrations caused by heavy footfalls.

If you find a pair of worms mating, you will notice one is facing one way, while the other faces the opposite direction. They grip each other with the long 'setae' on the belly and to help even more, the clitellum produces a thick sticky sheet of mucous which looks and functions a bit like sticky tape, keeping them close.

Over the next two or three hours the worms transfer sperm and go their separate ways. During the following few days each worm will produce more mucous and as it passes the male and female bits sperm and eggs are placed in it. As it rolls over the worm's head, the ends seal to form what for all the world looks like a small lemon – this is the worm's egg cocoon which now waits to hatch. From up to 20 eggs only a few baby worms will successfully hatch.

WORMS IN THE DARK

Like many nocturnal animals, worms cannot see red light, so cover torches with red cellophane if you want to watch them after dark.

You can make a study of worm behaviour with a simple but clever device calle d a 'wormery', an essential bit of kit for the oligochaetologist (someone who studies worms).

YOU WILL NEED:
piece of 2 cm x 2 cm wood, 116 cm long • 14 small wood screws • 2 pieces of clear perspex (30 cm x 30 cm) • 30 cm strip of wood, 4 cm wide elastic bands • a selection of different coloured soils, (garden soil, sand, potting compost) • 2 pieces of black card/paper 30 cm x 30 cm • screwdriver • drill • scissors • water • 5–6 worms

1: Cut the wood to lengths: 1 x 30 cm, 2 x 28 cm and 2 x 15 cm. Then drill two holes a few cm from the corners on each 30 cm piece of wood (these will eventually act as drainage holes) and a hole in each of the remaining 15 cm pieces, which will eventually become the feet of the wormery.

Using the 14 screws, screw the whole contraption together as in the illustration. The wood should be sandwiched between the two pieces of perspex, flush at the edges. The feet should be screwed on last.

2–4: Add the soils, one at a time, alternating layers, add a few leaves at the top and water lightly.

5: Add 5–6 worms and then place the 30 cm strip (4 cm wide) of wood on top, using the elastic bands to hold it in place. This acts as a lid and stops the worms escaping.

Then place the two pieces of paper or card over the sides of the wormery and keep in a cool place. Remove the paper/card blinds when you want to observe the worms within.

Check daily to make sure the soil is damp, but never soggy and wet, adding water if necessary.

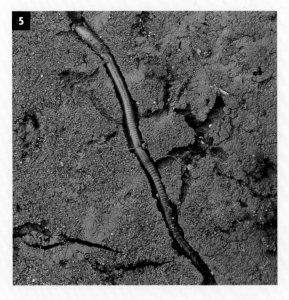

WORM CIGARS

1 Gingerly and usually under the cover of darkness, the worm quests for its prize, keeping its tail anchored in its burrow.

2 Having located a freshly fallen leaf, the worm grasps the stem in its mouth and drags the leaf back toward the burrow.

3 The leaf is folded and rolled into a cigar shape as the worm struggles to pull it down through its front door.

4 The leaf is often left to decompose a little before the worm tucks it into its burrow to eat. Nutrients in the leaf return to the soil.

What worms eat

Soil can be split into two main ingredients. First is the hard stuff, which is made up of rock that has been worn down by weather and water to form lots of tiny particles. The second ingredient is made from plants and animals that have rotted down. This is the magic stuff called humus, and it is humus that holds water, influences the texture of the soil and provides a lot of the nutrients that enable plants to grow.

Worms swallow soil as they move along, feeding on the humus and removing what they need. What comes out the other end is broken down even further so worms speed up the process of rotting or decomposition.

Have a close look at the surface of your lawn and you may notice what looks like hand-rolled cigars stuffed into tiny holes in the lawn. You may see that surrounding some of these 'cigars' are worm casts.

If you are lucky you might get to see some of these cigars being made. Go out on a warm, wet autumn night and scour the lawn for worm action, armed with a torch covered with red

Above: In autumn, look for earthworms' neatly rolled leaf 'cigars' stuffed into the surface of lawns.

cellophane (remember worms are blind to red light).

You may find worms mating or if you are really lucky you may catch a worm reaching out of its burrow, grabbing a leaf in its mouth and pulling it by its stem. The leaf will naturally fold and roll to fit the hole. The worm covers it with digestive spit and waits until it has softened before starting to eat it. By doing this, the worm is fertilising the soil and speeding up the composting process of the leaf.

This is why raking up leaves and tidying the lawn in autumn is bad for worms and bad for gardens. If you left the leaves long enough, most would end up buried in the lawn – think of all the raking and sweeping that would save!

Cast away

What exactly are those strange wiggly things that spring up all over lawns? There is no polite way of saying it, it's worm poo. To be precise it is the excrement of one of two species, *Allolobophora longa* and *A. nocturna*, which between them are almost entirely responsible for all the worm

casts that appear on the surface. Other species produce them, but they normally remain in the entrance of the burrow.

Don't worry about all this worm dung, it is not very unpleasant. Take some and rub it between your fingers and you will notice it is made of lots of tiny particles of soil. In fact that is pretty much all it is.

The great thing about worms is that they are so inefficient at digesting that if you were to analyse the contents of a worm cast you would find it contains 65–70% humus. This is good news because as it passes through the worm's body it is broken down into smaller pieces and a lot of the goodness it contains is now great fertiliser for plants.

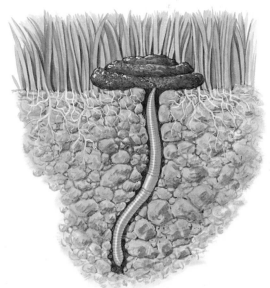

Above: In one end and out the other. As the worm tunnels it swallows the soil, digesting and absorbing what it needs, and leaving a worm cast.

Above: The best poo in the world? Worm casts are not at all unpleasant. They are partially digested soil, full of goodness for the garden.

Free gardening service

Some people hate worms messing up their lawn with casts, so they poison the soil and kill all the worms. Not only is it bad news for the worms but now the gardener is committed to a life of hard work. To stop the grass dying the gardener has to do the same work the earthworms were doing before. He has to spend a lot of time spiking the lawn to allow water to drain away and air to get to the grass roots. He also has to spend time and money adding fertiliser to the grass, to replace the nutrients that are washed away by the rain. Before, the earthworms would have brought these same nutrients to the surface in their worm casts. So leave those earthworms alone!

What eats worms?

Worms are ideal food – small, no tough chewy bits, easy to digest, high in protein and, if you know where to look, there are plenty of them.

Not surprisingly, other animals have sussed this out. Lots of birds enjoy worms and Hedgehogs, Moles and Badgers eat so many that they deserve the title of 'wormivore'. Slugs and some leeches will eat them and fishermen hang millions of them on fishing hooks every year to tempt a trout or two.

Believe it or not some people enjoy sucking on an annelid! If you want recipes, have a look at *The Worm Book* by Loren Nancarrow and Janet Hogan Taylor.

COMMON WORMS

Common Earthworm

My favourite. The biggest in the UK (up to 30 cm long), it can dig down over a metre. This is the one you're most likely to see on the lawn on a warm humid night.

Allolobophora longa

Despite the fact that it doesn't have a common name, this worm is very common indeed. It grows to 10-12 cm and is famous for its squiggly worm casts.

Angler's Red Worm

Looks a bit like the Common Earthworm but smaller and redder. These wriggle like mad when exposed to light, so fishermen love them for their ability to attract fish.

Brandling Worm

These stripy little animals are also called Tiger Worms. They can cope with big changes in temperature, acidity and wetness, so they're not too fussy where they live, from piles of leaves to compost heaps.

Slugs and snails

Slugs and snails belong to a huge group of animals known as molluscs, which includes clams, mussels, chitons, octopuses and squid. They belong to a class known as the gastropods, which is a cool name that means 'belly foot'.

Super slimies

Slugs and snails are not a very popular group of creepy-crawlies. This is partly because many people think they all want our lettuce and strawberries and have a penchant for our pansies. They are also slimy, which doesn't do much for their appeal.

But get to know them and they are a fascinating bunch. Of the 80 types of snail and 20 slugs in the UK, very few species actually cause damage to our gardens and some are even quite pretty.

The only real difference between slugs and snails is that snails have a hard shell that they can withdraw into while slugs do not. However, some slugs do have tiny shells on their backs and most species have a little one inside their bodies. And a few snails do not fit inside their own shells.

All slugs and snails are quite slimy. The slime slows down the rate at which water is lost from the surface of their skin. Because slugs do not have a shell they have much thicker slime to stop them drying out. Even so, some tough slugs and snails can lose around 50% of the water in their bodies and still survive.

Multi-purpose mantle

The mantle is the part of the body that secretes the shell in snails and forms a curtain that the snail hides behind when it retreats into its own shell. It can also produce a screen of bubbles that are blown at any intruder or curious conchologist (someone who studies slugs and snails).

Survival shells

Snails' shells protect them from predators and make them a difficult mouthful to swallow. They also act as a kind of survival capsule.

Slugs and snails lose a lot of water every day through their skin and the production of slime uses up a fair bit of water too, so they need to conserve it. Snails use their shell to help them do

Left: Face to face with a snail. There's definitely something great about going around with your home on your back!

Body design of slugs and snails

Shell The shell of a snail is instantly recognisable, even when the animal is long dead and gone.

Eyes

Tactile, tasting, tentacles All land slugs and snails have four tentacles that stick out of the head region.

Tail

Head

Mantle This is the the hump that contains all the gubbins of a slug, and the thicker, tough, rubbery flesh that can be seen around the mouth of the shell in a snail.

Glands on the skin The bit of the slug or snail that is exposed to the outside world is covered in tiny glands that secrete a thick mucous. This slime runs and oozes along a network of channels (which you can see if you use a microscope).

Breathing pore This is always found on the right of the mantle. It can be opened and closed by the snail, which is why sometimes you can look for it and not find anything!

this. Slugs do not have this advantage and that is the main reason why you can find snails in drier habitats than slugs.

When conditions get very cold in winter or extremely dry in summer snails can retreat into their shells and seal off the entrance with a thick mucous, which dries to form a waterproof seal called an epiphragm.

Wrinkles and rings

If you look at a snail's shell closely enough you will notice little ridges and sometimes a slightly different colour to sections of the shell.

Snails grow by adding new material to the mouth of the shell, which is formed with two layers of building material. The outer layer appears papery and thin at the lip of the shell as it starts growing in early spring. It acts like a waterproof varnish for the tough chalky layer that grows underneath. In a similar way to trees, snails have good times, where food is plentiful and they grow well, followed by hard times like winter or a really dry spell when the snail hardly grows at all. The ridges and wrinkles in the shell are formed when conditions change.

Breathe with ease

The breathing pore is a hole leading to a chamber called the mantle cavity, which is used a little bit like a lung. Air wafts in and oxygen is taken up by a network of blood vessels in the skin. Some pond snails use this technique too. If you keep a few Great Pond Snails in a jam jar next to your bed at night you can actually hear them breathing, making a kind of popping noise as they come to the surface and open their breathing pore.

Some other water snails have gills and fill their mantle cavity with water, from which they extract oxygen. Others have a combination of both. All slugs and snails can also breathe through their skin.

Tentacles: tactile tasters

Look closely and you will see a dark spot right at the very tip of the tentacles. This is the eye, but it can't see very much. (About as much as you could see if you stuck greaseproof paper over your own eyes. You could see the difference between light and dark but that's about all!)

Snails are very sensitive to light and they follow the low light of the night out into the garden to feed. Tentacles are also used to feel around, just like you might use your hands if you were blindfolded. You can steer a slug or snail by gently touching the tentacles one side at a time.

Each of the four tentacles is covered in taste buds. The lower ones are sensitive to food up to 20 cm away, while the two larger top tentacles work over distances of 50 cm or more.

Because they are so important, the tentacles can be sucked back into the body, in the same way a glove finger can be turned inside out. Gently tap a tentacle with your finger and watch as it rolls in on itself. When the coast is clear the mollusc squeezes blood back into its tentacles and they roll out again. If they do get damaged it still is not the end of the world as new ones grow to replace them.

Making more molluscs

Like the worms in the previous chapter, there is no such thing as a he or she snail or slug. Each animal is both, which is helpful for such famously slow-moving animals. If they do bump into each other they make the most of it! Both partners will be fertilised, so both can lay eggs.

Take a torch out on a damp and warm night— you never know what you might see. Snail and slug love can be as simple as a kind of mollusc kiss chase, with one animal picking up the trail of the one it fancies and following it until the potential mate is found. After a quick embrace both animals go on their way having exchanged sperm via a pore on the side of their bodies.

Cupid's arrow?

Certain snails have a surprisingly passionate courtship. When a Garden Snail encounters a mate they meet each other head on, rear up and with their mouths pressed together they 'kiss'. Then they fire love darts at each other. (I'm not kidding – such passion!) After stabbing each other with these sharp shards of shell-like material, they mate. Nobody is really sure what these darts do; they could stimulate the other snail into producing sperm or they could inhibit the other snail from mating with another.

Egg fest

If you poke around in the soil and explore beneath bark and stones, sooner or later you will stumble upon what looks like a cluster of mini ping pong balls. These are the eggs of slugs and snails, which are laid in damp crevices so they won't dry out. The size of the clutch varies a lot, but anywhere between 10 and 100 eggs is common.

It is fun to collect these eggs and watch them hatch, which will take a few weeks, depending on the temperature. You will see the little molluscs inside the egg just before they break free.

The eggs of pond snails are very different and have a jelly-like coating for protection. They are usually found on pond weed or on the undersides of lily pads and stone surfaces.

Munching molluscs

A lot of people will throw down slug pellets and slowly turn the subject of this chapter into a slimy lifeless

SPECTACULARLY SEXY SLUG!

The Great Grey Slug has some of the most spectacular sex in the animal world and if you are ever fortunate enough to witness it you will feel like giving them a round of applause when they have finished. It goes a bit like this.

First the slugs meet, they run each other around and work up a bit of a lather, producing loads of mucous, then one slug initiates the act of 'going upstairs'. They climb up a vertical surface, still chasing each other. When they find a mutually acceptable spot for their affair they start tickling each other with their tentacles. As they get more and more involved with each other they continue to ooze slime and further entwine until they make the ultimate lovers' leap!

blob. This may get rid of the nightly nibblers of prized seedlings, but for every slug maliciously murdered in this way, I bet there are several other animals going hungry that we would rather like to see in our gardens.

Not every animal hates slugs: some positively love them. And remember that any slugs and snails that are slowly dying can be consumed by other animals. By using chemicals and poisons on any pest, you could also be poisoning animals you would love to have in your garden.

One of the top mollusc mashers leaves vital clues at the scene of the crime. Shards and splinters of snail shells scattered around prominent rocks, stones, paths and patios are the work of the Song Thrush. This bird uses 'anvils' to dash the snail's shell to pieces and get to the succulent morsel inside. Both Mistle Thrush and Blackbirds also put away large numbers of slugs and snails. They just haven't invented such an effective way to get the

Left: The eggs of the Garden Snail. I often stumble upon clusters like this while digging the flowerbed or lifting stones and logs.

wrappers off.

Look under corrugated tin sheets, often the home of Field and Bank Voles. Here you will find neatly piled clusters of snail shells, stacked like broken crockery. Shrews and Hedgehogs will also snack on snails. Toads also provide a great slug removal service, as does the Slow Worm.

Ghastly slug guzzlers!

All of the above slug guzzlers will eat the animals whole, but the last thing a slug wants to see is a Ground Beetle. These have vicious mandibles that can slice open a slug or crunch through a snail shell like a pair of bolt cutters.

Even worse is a Glow Worm larva. These unusual beetles prey almost entirely on slugs and snails. They repeatedly stab their mollusc victim with venom-laden mouthparts, and when it is not able to run away or even twitch, the larvae eat the animal alive.

GARLIC SNAIL

The Garlic Snail, as its name suggests, tries to put off its attacker by smelling strongly of garlic!

In their defence...

Slugs and snails can defend themselves pretty well from all these attackers. The obvious defence for a snail is to withdraw its soft parts into its protective shell. If you pick up a snail and poke it, eventually it will belch forth an impressive quantity of bubbles. This sticky green froth gets everywhere and will certainly put off many predators that might otherwise crunch the shell.

Slugs, despite appearing quite vulnerable without a shell, are not as defenceless as you might at first think. They can hunch up into a ball shape in seconds, withdrawing their tentacles and exposing large hump of its mantle, which is covered in a thicker leathery skin.

This posture also has the effect of making the slug harder to pick up or bite into, especially

Above: No prizes for table manners for the Song Thrush. It beats snail shells against a favourite rock or 'anvil' until they shatter.

when it starts to produce a thicker slime in its defence. Try picking one up and you will find the goo is like glue.

I once saw a Blackbird pick up a Great Black Slug but it soon lost interest and spent the next five minutes trying to wipe the mucous off its beak.

This same species of slug also starts rocking when it is threatened. Quite how or why this is scary I'm not sure but it certainly looks fairly funny if you get a chance to see it happening.

Below: Great Pond Snails are common freshwater snails. They cannot breathe under water, a design fault perhaps for a water snail, but they seem to manage just fine returning to the surface to get a lung full of air.

YOU WILL NEED:

**small piece of clear glass or perspex • a slug or snail •
misting spray of water • magnifying hand len**

Take a snail or a slug – it doesn't matter which.
If you are not too keen to touch them you can use a
couple of spoons to pick up most species and avoid
getting sticky fingers. Place your chosen animal onto the
piece of horizontal perspex or glass and wait for it to
start moving. This sounds easier than it actually is.
If after a few minutes your gastropod refuses to emerge
from its shell, or even shake a tentacle,
it is time for you to get persuasive!

Most slugs and snails behave like this when the
air is too dry for them. You can reassure them that
conditions are good by lightly spraying them with water.
(Run your fingers through a wet nail brush
to produce a fine mist of water if you don't have a
spray-mister.)

When your mollusc is moving gently lift up the clear
sheet and watch from below. You should see dark and
light bands moving along the underside of the foot.
These are bands of muscles, the dark ones are raised and
'stepping' forward, the pale bands are in contact with the
surface. But the mollusc wouldn't get anywhere without
the essential ingredient – slime. Glands on the underside
of the snail produce loads of this stuff, which lubricates
the ground and allows the mollusc almost to surf along
though the garden.

On the snail trail

The first bit of this next experiment needs you
to be a top tracker. You have to find a 'roost' of
snails. Garden Snails are the best, because they
are big, darkish in colour and common. Think
about what a snail needs: shelter from predators
and the sun. Look under piles of rocks and logs,
walls, flower pots, overhanging plants, even ivy.

Once you have found your snails, make a note
of where they are and then using enamel paint,
mark each individual snail's shell with a number,
making sure you do not get any paint on their
soft parts.

Return the next day and see how many are
still there. Mark any new ones. You can even look
for other day-time hide-outs in the same garden
or park and mark these with another colour. Do

Below: Slugs and snails can be persuaded to reveal their
slippery secrets by placing them on a sheet of glass.
Look for the dark bands of moving muscle.

the snails mix up or do they return to the same place? If you do this well, soon you will know all the snails in your own patch personally, where they live and how many there are. If you are really keen, you can go out at night with a torch and look for them, plotting their positions on a map so you can work out how far they travel during the night.

SOME SNAILS AND SLUGS

Garden Snail A very common snail, second only to the rare Roman Snail in size. Its large appetite for garden plants makes it unpopular but it is beautiful to anyone prepared to look at it the right way.

Strawberry Snail A very common small snail found all around gardens and in greenhouses.

Door Snail Many snails have a corkscrew swirl to their shells. The Door Snail is so called because of a 'door like' mechanism that shuts the shell.

Great Pond Snail The commonest water snail in areas of hard water, found in ponds, ditches and lakes.

Great Black Slug This handsome beast is usualy glossy black, but it also comes in a bright orange form. A fully grown one will reach 15 cm in length.

Hairy Snail Yes, snails can be fluffy!

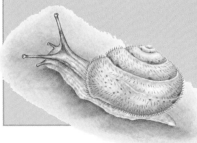

Ramshorn Snail One of a large group of molluscs with flattened shells, which look a bit like the coiled horns of male sheep.

Shelled Slug Quite common, but rarely seen, this slug spends most of its time underground chasing worms and other slugs. It has a tiny shell.

YOU WILL NEED:

blender • lettuce or grass or cornstarch • cuttlefish bone (from a pet shop) • small piece of perspex or glass • paint brush • snail

There are lots of different ways of seeing a slug or snail's radula, but you will need a good magnifying lens and a cooperative snail to get a really good look. The best chance is with our old friend the Garden Snail. Not only has it got a big mouth, it will eat almost anything.

First of all you need to tempt your snail to eat by making a tasty soup. I've tried lettuce, which works well, but you could experiment with other greens. Place the lettuce in a blender and turn it into a runny liquid, add a bit of water and a bit of chalk in the form of cuttle bone (snails need the calcium for their shells).

Using the paint brush, paint this liquid onto one side of the clear perspex or glass and leave it to dry. Repeat this a couple of times to build up a thicker layer.

Have a heart

If you grow pond snails in a small aquarium or tub that is kept under bright light and make sure they are fed well, they tend to grow quickly and their shells tend to be thin. This is handy for the snail enthusiast. Use a magnifying lens to see into the shell of a living snail and you will notice the mantle cavity and beating away next to it the two chambers of the heart.

Below: Banded snails are quite conspicuous. The White-lipped and Brown-lipped Snails are variable. They can be banded dark brown or black on yellow, the bands can be missing altogether or can be brown!

Woodlice

All the animals mentioned so far – worms, slugs and snails – do not have legs. The rest of the animals in the book do have legs. They are all arthropods (which means jointed legs).

Suits of armour

All arthropods have legs and bodies that have hard bits on the outside for at least part of their lifecycle. They need joints to move about – a bit like suits of armour. Imagine wearing one all the time, though – as you grew up, it would get tighter and tighter and would eventually start to hurt. If you kept it on, you would end up permanently damaged and deformed.

Arthropods have the same problem and they solve it in the same way that you would – they climb out of their suit and get one the next size up. The clever bit is that they have a soft suit underneath their old one. It's a bit baggy to start with but after it is pumped full of blood and air it soon hardens up.

These animals have no backbone and no skeleton whatsoever inside their bodies. But the armour, which is called an exoskeleton, is like a skeleton on the outside.

In the same way that our skeleton provides places to attach muscles and tendons, so it is with arthropods, just inside out.

Many arthropods have especially tough exoskeletons, some with knobbles, spikes and barbs on as extra protection and defence.

Mini robots

With a bit of added wax and oil a skeleton can be waterproof on the outside too. This waterproof skin actually keeps water in, which means the animal can live in dry and hot places.

Lots of useful things can be found added to exoskeletons – like a spring-tail's spring, extra eyes, a beetle's mouthparts and even wings. Arthropods are a bit like little robots with lots of accessories for their own way of life. The big disadvantage is that when they climb out of their old skeleton to inflate their new one arthropods become vulnerable to both predators and to the elements.

Left: The wonderful woodlouse, complete with armour plating.

MOULTING

Most other arthropods shed their suit of armour all at once, but woodlice do it in two stages. The back end comes off first and the new exoskeleton is left to harden, then the front end is moulted. If you discover woodlice mid-moult they do look a little odd. The soft bit is always a pale, almost white, colour, giving you a two-tone woodlouse. The colour, shape and texture of these plates are the best way to tell most species of woodlice apart.

What are woodlice?

Woodlice have too many legs to be insects and not enough to be a millipede or centipede. So what exactly are these strange little beasts?

Scientists have placed woodlice in a group called the isopods, which means 'identical legs'. All the walking legs are about the same length and look almost identical. Isopods are crustaceans, a group that includes lobsters, crabs and shrimps.

If you catch a woodlouse and gently flip it over you will see seven pairs of scrabbling legs and a small pale area behind the last pair of legs. These are gills, structures which absorb oxygen from water and get rid of waste gases from the bloodstream.

The gills of water crustaceans, like prawns and freshwater shrimp, are exposed – you can see them flapping about in the water.

Woodlice gills don't flap about like this, but they still need water to work properly, which is one of the reasons woodlice are found only in wet and damp places – they need water to breathe. If you look at an upturned woodlouse with a magnifying lens of at least x15, you will be able to see water sloshing around in its body.

Woodlice bodies are protected by lots of tough overlapping plates. They are very hard but they are not very waterproof, so woodlice can dry out very quickly. Some species, such as the Pill Woodlouse, survive better than others in dry habitats because their plates are covered with a waxy layer that stops water passing out of their bodies.

Broody baggage

Sometimes you can come across a female woodlouse with a brood pouch between her front legs (visible on the underside). This sometimes looks white or yellow and it is here that the eggs are kept until they hatch out into a pool of liquid. The babies remain in this pouch until they are big enough to survive in the outside world.

The same ... but different

Woodlice can be found in all kinds of places. They are nocturnal (active at night) but you don't have to look very hard to find their daytime hideouts.

I used to get into trouble with my mum as I was always disturbing her rockery to find them, but you can find them under almost any garden object that provides some dark and damp shelter. Try looking under logs, stones, bricks or bits of discarded tin.

Forage around for long enough and you will notice that there are lots of different types of woodlice. Some only differ in the number of segments in their antennae, others have a completely different colour or texture. Once you have noticed this you might find this group of animals rather addictive.

Unlucky for some

Nowadays many people think of woodlice as pests, especially if they see one in the house. But in the past they were revered and respected and even featured in paintings and poetry. They have been eaten by humans and used as medicine to

The woodlouse body

Uropods The back end of a woodlouse has four little pointy bits sticking out. These are called uropods. There are actually two of them but they fork to give the impression of four. They are like rear antennae, and may release nasty chemicals to put off sneaky predators trying to creep up from behind.

Hard armour Tough plates made of calcium carbonate.

Antennae Used for feeling their way in dark, dank places.

Legs Fourteen legs are used for walking.

Eyes A cluster of simple eyes called ocelli is found on either side of the head in most species.

Gills Turn a woodlouse over and you'll see pale gills at the back end.

Learn a little about woodlice biology, their social life and how they got their German name of 'pissibeds'!

YOU WILL NEED:
small jam jar with lid • some leaves and bark • damp soil • selection of woodlice

1: First make a nice little home for your woodlice. Place damp soil and a few leaves and bark into your jar.

2: Pop your woodlice in the jar and put the lid on. Leave the jar in a cool dark place for a couple of days. There will be enough air in the jar to last them for some time.

3: After a few days, quickly remove the lid and take a big sniff from the jar. Pheweee! What a pong! It smells of wee.

4: Set your woodlice free.

cure illnesses from indigestion and ulcers to tuberculosis. In fact they were often worn around the neck in a little locket to be taken like a kind of primitive indigestion tablet! Cows were also given them to help them chew the cud, hence one of their other names, 'cudworm'.

I don't know why woodlice are considered pests just because they are creepy-crawlies. Some people consider it unlucky to have woodlice in their home. Well, when a home is damp woodlice feel at home too, so I guess you are pretty unlucky if you have to fork out for new double glazing or a damp course for your home. But it's not the woodlice's fault!

Top class composters

Woodlice rarely eat living plants; they might nibble at particularly succulent and tender seedlings but nothing serious. Most of their diet is dead or rotten material and they will also graze algae and have a munch of fungi too. They play an important role in the world of composting, breaking down bits of dead plant in garden compost heaps and recycling dead wood and leaf litter into the soil.

Weird friends, cool enemies!

Woodlice are quite long-lived. They can live to the ripe old age of four, meaning they have the same kind of life expectancy as a gerbil! However, it isn't exactly a stress-free existence for them in the woodlouse world.

Despite all their tank-like armour and the nasty secretions produced by their bodies for defence purposes, woodlice are prey to many creatures. It is surprising how many are partial to a Porcellio (woodlouse) from time to time.

Apparently, 40% of all woodlice are eaten by centipedes, whose sharp, sickle-like jaws can make short work of an armour-plated louse.

Shrews and Hedgehogs will put 'em away given half a chance and toads have been reared almost entirely on a diet of woodlice with no apparent ill effects.

There are a few specialist 'hit creatures' that will prey entirely on these unfortunate crustaceans. They present particularly unpleasant ways to go. How would you choose between death by Dysdera the spider and being eaten alive by the maggots of a parasitic fly?

ATTRACTING WOODLICE TO YOUR GARDEN

There are about 37 different kind of woodlouse found in the British Isles today. Some 30 or so are native, the rest have been introduced from abroad. If you want to encourage these little suits of armour into your garden, go and getsome dead wood and make a log pile in a corner. The most important thing is to leave the wood alone; if you expose them too often they will move on. the wood alone; if you expose them too often they will move on.

Right: Woodlice often cluster in dank dark places. They can't tolerate dry conditions for long because they need water to live and breathe.

Experiment: Give 'em a choice

Woodlice movements are controlled by light and humidity (the amount of moisture in the air). They will keep moving if things feel uncomfortable, so a woodlouse out of cover on a sunny day will simply keep on walking until it finds its ideal habitat.

Try these simple experiments to find out exactly what makes a woodlouse tick and why you find them where you do..

YOU WILL NEED:

board 30 cm x 30 cm • plasticine • sheet of perspex 30 cm x 30 cm • black card or paper • cotton wool or toilet paper • water • water-based marker pen

1: Place your board onto a table or a flat surface.

2: Roll out your plasticine to form long sausage shapes and make a wall all around the edges of the board. This is the arena for your experiment.

3: Divide the arena with three smaller strips of plasticine, leaving two small gaps for the woodlice to get through.

4: Place the perspex on top like a roof.

5: Now you are ready to follow the two parts of the experiment.

PART 1: WET OR DRY?

Remove the perspex and place a piece of damp cotton wool or paper in one side of the arena. Put five woodlice in each compartment and replace the roof. Watch and see where the woodlice end up after a couple of hours. Do they like it moist or dry?

You can add another dimension to this experiment by repeating it with just one woodlouse. Place it in the side without the cotton wool and use your marker pen to trace its movements on the perspex. What happens to its movements when it steps into the damp and humid side of the arena?

PART 2: LIGHT OR DARK?

Remove the damp cotton wool and cover one of the compartments with black card or paper. Put five woodlice in each compartment and put back the perspex roof. Wait for a couple of hours. Where are all the woodlice now?

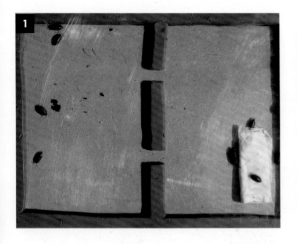

SOME COMMON WOODLICE

Porcellio scaber

Very common in gardens under stones or logs, but may also be found in trees.

Pill Woodlouse

Also called pill bugs, these are often found at the base of walls. They roll themselves into a ball if you disturb them.

Androniscus dentiger

Only 6 mm long and very pale. Lives in compost heaps, cellars and leaf litter.

Common Woodlouse

Flatter than other woodlice. Found in clusters under stones, logs and bark.

Centipedes and millipedes are myriapods, which means 'many legs'. But not that many: millipedes do not have thousands of legs and centipedes certainly don't have hundreds! Although they have similar body designs, they are not actually closely related.

The secret's in the segments

Both animals have a body that is made up of jointed body, a bit like a bicycle chain. This means they can have all the advantages of a tough outer skeleton plus the flexibility of a worm, allowing them to move easily among leaf litter and soil.

Sensitive centipedes

There are many different species of centipede but they are all predators, eating other small creatures in the soil or leaf litter. They are the stuff of nightmares for tiny creatures that like to live under logs.

Not only are they fast, but thanks to their long legs and flat body they can chase down their quarry even when it tries to hide in a nook or cranny.

All these qualities make centipedes very easy to recognise so if you find an animal with lots of legs, which starts sprinting around, you can be sure you've got yourself a centipede!

Of the 50 or so centipede species that live in Britain, all belong to two basic body designs. Over half are long and stringy and live in the soil. They are very narrow with very long bodies. They are so flexible that they can tie their own bodies in knots and will often be in this tied-up position when you dig them up. Most of these soil-dwelling centipedes are an anaemic-looking yellow or white.

The other type are flatter and live under stones and logs and among the leaf litter. These have fewer legs than the soil-dwellers, but are much bigger and bolder.

Pinchers and feelers

Centipedes are particularly effective predators because of the two evil-looking claws underneath the head. These are actually hollow legs equipped with a venom gland. When centipedes catch their prey they give it a quick nip and inject venom, which will quickly kill even the most energetic woodlouse. I strongly recommend you

IT'S A RECORD

The world record for most legs belongs to a millipede with 750.

Left: Lift stones, dig in the soil or look under bark and you're bound to find centipedes and millipedes.

look at one of these awesome animals with a magnifying lens – if you can slow one down for long enough! Get one in your bug restrainer if you can, and try looking for the eyes. You should spot the tiny simple eyes on some species, positioned low down on the side of the head. These little eyes are not much use, so how does a centipede catch its prey and get around – especially in the dark?

Well, the clue is in the question. It is so dark in their world that centipedes don't actually need eyes. Instead they feel their way with a splendid set of antennae. These are very sensitive to vibration and touch and probably to smell too. At the other end of their body are modified legs which also act as feelers. In fact, every leg is sensitive. No wonder that centipedes are so highly strung, a fact that becomes apparent if you try to catch one and put it in the bug restrainer!

Mooching millipedes

Millipedes feed on very different stuff to centipedes. In fact, they are are mostly vegetarian. Some scavenge, but none of them actually feed on living animals. And because plants and dead things do not move very fast, millipedes don't have to either. They are very chilled out compared to centipedes!

Smooth operator

Most millipedes have a very different body shape to centipedes, with a chunky profile and four short legs attached to each segment.

MARVELLOUS MILLIPEDES

Snake Millipede
These have a round profile, like a vacuum cleaner hose. They produce really nasty chemicals, including cyanide, which makes them very unpleasant to eat. (They are not dangerous to humans.)

Pill Millipede
Often mistaken for the similar-looking Pill Woodlouse, but check out the legs: a Pill Millipede has loads more legs than the woodlouse. They are very good at avoiding desiccation by curling up into a tight ball, so they can cope with drier conditions than other millipedes.

Flat-backed Millipede
Also known as Polydesmids, thesey tube-shaped millipedes have shield-like extensions sticking out either side of their body. You may catch a whiff of almonds if you have one captive in a pot - like other millipedes they defend themselves with chemicals and this is the smell of cyanide.

The long smooth body allows millipedes to fit into narrow crevices and they can generate a huge pushing force with their many legs. (Try holding a millipede in a loosely clenched fist and you will feel how strong it is as it tries to push its way between your fingers.) Because they do not have big flouncy feelers and long legs like a centipede they have nothing that is likely to get caught on things and they can barge their way through soil and leaf litter. Like centipedes, they have poor eyesight, but they do not need huge feelers because their food doesn't run around. Instead, they make do with a set of short antennae on the head.

Centipede or millipede?

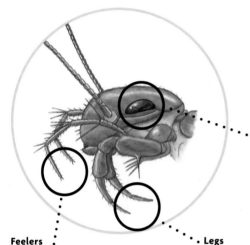

CENTIPEDE Because of their fast pace of life centipedes are bristling with sensory equipment to help them cope. They have a flat profile for slipping into crevices and between the leaf litter.

Simple eyes
Because centipedes live in dark places their eyes do not need to be that advanced, just good enough to tell dark from light.

Feelers
Long flexible antennae for sensing their prey and environment.

Legs
Modified front legs to pinch prey and inject venom.

Simple eyes
Clusters of simple eyes are set low on the head. They are even more basic than a centipede's.

MILLIPEDE Compared with the centipede these are slow and sluggish, designed to push through soil and litter. They have a strong and powerful tube-shaped profile.

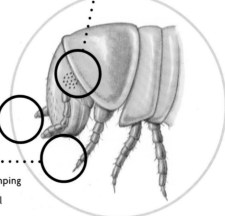

Feelers
The antennae are short and small and seem to taste and touch the ground in front of the millipede's head

Mouthparts
Strong and sturdy, for chomping through dead plant material

Arachnids

Let's get this straight – anything with more than six legs is not an insect. Arachnids are a group of mini-beasts that includes spiders and a few other oddballs. They come in a variety of shapes and sizes but they all have two things in common: four pairs of walking legs and no antennae.

Left: A Garden Cross Spider awaits unwary prey.

Super spiders

At times it seems the whole world is suffering from arachnophobia or what I call 'Miss Muffet Syndrome'. Just mention the word 'spider', let alone reveal a living one, and someone will almost certainly fly out of the room in hysterics!

Although all spiders use venom to subdue their prey and a few can just about manage to sink a fang through human skin, there is no reason to fear any British spider.

Spiders come in many shapes and sizes: gangly ones, dumpy ones, fat ones and thin ones. Some are very hairy, others appear smooth as silk, some are sombre in colour and others are actually quite pretty!

There are spiders that hunt alone, spiders that spin elaborate traps, underwater spiders, surfing spiders, parachuting spiders, spiders that jump, spiders that steal from other spiders and even pirate spiders that murder their fellows. They are a truly versatile bunch.

Know your way around a spider

With the naked eye you can see that a spider's body is in two parts joined by a narrow waist called a pedicel. For a more thorough examination, you will need a hand lens and it will help to use a bug restrainer for some of the faster ones.

The front bit is a bit like the head and thorax of an insect combined. Its technical name is cephalothorax, which means 'head chest'. The cephalothorax is the engine room and control centre of a spider. It contains a lot of the nervous system and muscles, the stomach and the venom glands, all encased in a box-like external skeleton, with a hard shield on top.

The parts of a spider

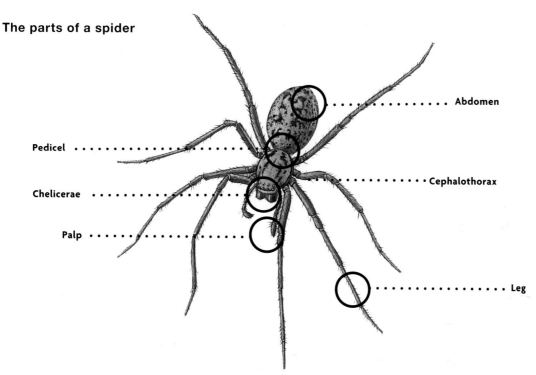

Abdomen

Pedicel

Cephalothorax

Chelicerae

Palp

Leg

Eyes

The eyes of most spiders are so simple that most cannot see very well at all. (Apart from some jumping spiders, which can see detail up to 30 cm away.) Some spiders definitely can use their eyes to detect mates and predators by seeing detail and movement. Some wolf spiders use light to navigate by and all spiders can at least detect differences in light and dark.

Most species have eight eyes arranged in two rows but there are always exceptions to the rules – some have six and some have only four. Others have a large growth with eyes placed directly on top of the spider's head. The differences are very useful when identifying spiders.

Right: The eyes of a jumping spider, some of the best in the arthropod world.

Tools of the trade

The 'mouth' of a spider is a set of structures that process the unfortunate prey in stages.

The fangs – also called the chelicerae – are situated either side of the head. They come in two parts: you can see the bases below a spider's eyes if you look at it head on but the sharp tip is often tucked away. These are the nearest thing these animals have to hands – not only are they used on prey, they can also pick things up, wind up silk, dig burrows and carry eggs around. Some of the tiny money spiders even produce noises by rubbing them together!

Venom is squeezed out of the muscular venom gland in the head region, and injected into the prey through a microscopic hole in the fang, a bit like a hypodermic needle. The venom paralyses the prey to stop it struggling and then it slowly dissolves the prey's insides, creating a liquid lunch.

Spiders have a very muscular stomach attached to inside walls of the cephalothorax and they suck up this liquid, filtering it through several layers of hairs and bristles that stick out around the mouth.

Protruding palps

Everybody knows spiders have eight legs, but when you count them you might think it looks as though they have ten. Sticking out either side of the head are palps, which look like legs. But besides being one segment shorter than a leg, they are definitely not used for walking. If you watch a spider for a while you will see it use the palps to tap things as it travels. A spider's palps are super sensitive and are used to grab, feel and taste prey. They also play a big role in the sex lives of male spiders.

The next time you see a House Spider have a close look at its palps. If it is a male (and they usually are because it is the males that go 'walkies' looking for females), it will look as if it has boxing gloves on. These structures are used to inject sperm into the female. Some species also wave them about in courtship or threat displays.

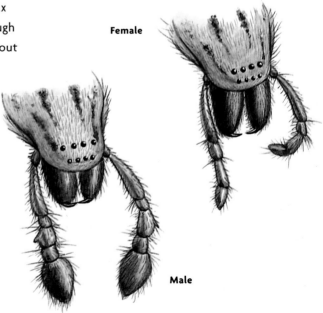

Female

Male

Right: He or she? Many male spiders have swollen tips to their palps, used in courtship displays and as sexual pipettes to inject sperm into the female.

Why are spiders hairy?

Spiders' hairy legs are not just to make them look scary. A bald spider would be a dead one as spiders need bristles for all manner of essential tasks. To a spider its hairs are camouflage, ears, nose, grippers, tasters, comb and hairbrush! There are many different kinds of bristles cut out for different kinds of work.

Some of the hairs that cover the body in a dense velvet give spiders their camouflage, patterns and sometimes groovy colours. The colours act to break up the spider's outline and help them blend into their background.

Fancy feet

It's hair that enables spiders to crawl up the smooth surfaces of walls and take a stroll across the ceiling. Each foot has a device made from lots of special hairs, called scopula hairs (scopa means broom in Latin). Each one of these hairs looks just like a brush divided into hundreds of even smaller extensions like tiny bendy spades. Each 'spade' uses the surface tension of the microscopic film of water found on most objects – a bit like the effect when a coaster sticks to the bottom of a wet glass.

Even a spider with only a few of these hairs on its foot, such as a Crab Spider which has only 30, can achieve 160,000 separate grips on a surface. This is why species with even more scopula hairs than this can support more than ten times their own weight when resting on even the smoothest glass surface. Other hairs on the feet are sensitive to tastes and chemicals and on the legs there are hairs used to comb the silk and yet others are used as a brush to keep all the other hairs in good working order.

Good vibrations

Spiders haven't got ears but they can pick up vibrations and sound waves in the air which is pretty much what our ears do. Just about all the hairs and bristles scattered on their body have a nerve ending. These allow the spiders to feel their environment with their whole body all the time.

The larger hairs are mainly used as triggers to tell a spider that something is moving or not – watch how the spider reacts if you try tickling one of these hairs with a pin. The finer, feathery hairs (called trichobothria) are easily 'blown' by even a tiny vibration or breeze and can tell a spider where its next meal is coming from.

Just the job

Some spiders have specialised hairs for specialised jobs. Water spiders rely on the hairs and bristles on their body to trap a layer of air in place, allowing them to breathe under water.

Wolf spiders can carry all their babies on their backs because their hairs are modified with a knob which acts a bit like a handle or grip for their young to hold on to.

Above: Using a webbed tunnel and the highly sensitive bristles on its body, this Labyrinth Spider has efficiently detected and despatched a cricket.

Breathing with books?

Most of the time spiders are slow unless running from a predator or pouncing on prey, when of course speed is of the essence. They can't keep their speed up for long though, because they do not have efficient lungs.

They have a special organ called a 'book lung' – it's not easy to see but you may just make out a tiny pair of slits on the underside of the abdomen, near the waist. Air passes into the spider here and wafts over lots of thin plates that look like the pages of a book. From here oxygen passes to the blood and at the same time waste gases pass out.

More active spiders may also have a tiny hole in the body wall, called a spiracle, which leads to a network of fine tubes that take air from outside to the internal organs.

Wandering wolf spiders

Walk through long grass in the summer and you may notice small spiders scatter from your path in huge numbers. Get down on your hands and knees and look at these spiders and you will notice a miracle in micro-mothering.

Most of these spiders will be wolf spiders, so called because they hunt like wolves. They use their big eyes to spot their prey and then they chase it down and pounce. They do not spin a web, and they don't hunt in packs, although sometimes it may appear that way because there are so many of them.

One of the commonest is a species called Pardosa amentata. Like many of the wolf spiders it is various shades of brown with a beige stripe running down its back, perfect camouflage for a hunter.

Because these spiders do not have a web, the females carry their eggs around with them in a silken cocoon slung beneath the abdomen. You might see the off-white egg sac before you see the spider carrying it.

If you peer a little closer you will see females carrying a bundle of spiderlings on their backs. Wolf spiders are good mums and will carry their young for about a week, giving them a head start before they wander off into the grassy jungle on their own.

Below: Spotted Wolf Spider. The female carries her eggs in a sac attached to her spinnerets.

The second half

Now to the back end of the spider: its abdomen. You can think of it as a kind of living 'water bomb' – lots of bits of plumbing, a heart, breathing apparatus, sex organs, guts and the all-important silk glands all contained within a sloshy blood bath.

The size of the abdomen varies. It can be podgy, plump and tight or withered and shrivelled, big and round or slim and slender, all depending on what species, what sex the spider is and at what stage of its life-cycle it's at.

Super silk

Silk is the spiders' multi-purpose secret weapon. They use it as a building material, safety line, trap, glue and parachute. Other invertebrates produce silk – caterpillars and caddisfly nymphs to name a couple – but none are as masterly in their use of it as spiders.

Imagine being able to produce a substance with which you could build a house, support many times your body weight, glue things together, catch your food and fly and at the end of the day you could roll it all up and recycle it! Sounds like the ideal super substance. The true silk masters, the orb web spinners, can produce up to eight different kinds of silk.

Silk comes out of the spider's body close to tip of its bottom. If you look closely at some species you will see tiny little projections that the animal seems to wiggle and wave – these are spinnerets. If you were to zoom in with a very powerful microscope you would see six spinnerets with lots of little nozzles or spigots. It is out of these spigots that the silk flows.

Look on dry banks and especially walls with crumbly mortar and you will probably see tiny white burrows, reinforced with a dense weave of silk. Look for the almost invisible radiating threads arranged like the spokes of a bicycle wheel.

The spiders down these burrows spend the majority of thier lives waiting in the security of the hole. The only way to tempt them out is to trick them into thinking you are dinner.

There are a couple of ways to do this. The first is handy if you happen to have a musician in the family as you'll need a tuning fork.

Strike the tuning fork and gently touch the 'trip' wires surrounding the spider's lair.

By doing this you are fooling the spider into thinking you are an insect and it will rush out from its hideout to attack.

If you have trouble getting a tuning fork, try a party blower with a piece of grass taped to the end. Unravel the tube, hold the grass against one of the threads and blow.

An alternative is simply tickling the web with a blade of grass, although this doesn't always fool the spider.

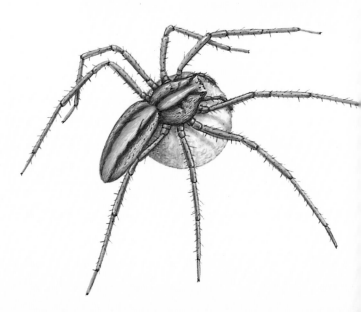

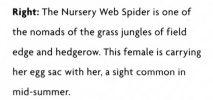

Right: The Nursery Web Spider is one of the nomads of the grass jungles of field edge and hedgerow. This female is carrying her egg sac with her, a sight common in mid-summer.

Silk starts off as a kind of protein soup within glands in the spider's body. As the silk comes out of the spigot, a complicated and rather magical process occurs and the silk stretches and hardens.

Watch a spider descending from the ceiling on a line and see what it does with its legs. It has one leg on the line all the time, controlling its rate of descent by pulling the silk out of the spigots, a bit like someone abseiling controls their rope.

Doilies of death

Think of a web and you probably have an image of the 'classic' orb web – those gorgeous wheel-like structures that are certainly the most elaborate of all the spiders' craft. These may look pretty but its true function is more sinister: to snare unfortunate insects. There are nearly as many different types of web as there are different spiders, ranging from simple sheets and tangles of silk to full-blown masterpieces.

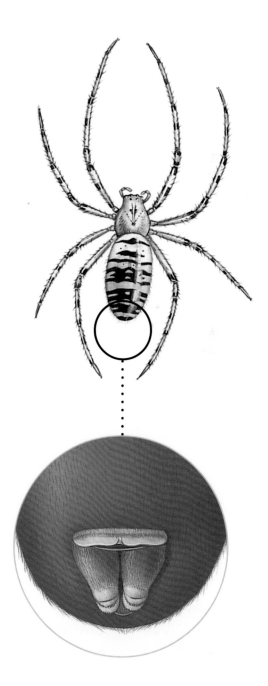

Right: If you turn a spider upside down in your bug restrainer and look at the end of its abdomen, you should see the spinnerets, the organs that produce and weave the silk that spiders are so famous for.

1: The spider has to span the gap. It does this by letting the silk drift on the breeze and by jumping or walking between the two points. Once the line is fixed and taut, the spider then crawls along its bridge, spinning a second stretchy loose silk strand. Having attached both ends, the spider then returns to the centre and pulls the second strand down to form the main skeleton.

2: More threads are added, to form the spokes of the web, anchoring it to the surrounding vegetation.

3: The spider then returns to the centre of the web and begins laying down the 'real' trap, the sticky stuff that actually catches the prey. Silk coated with a sticky liquid kind of protein is what makes up the spiral.

4: The web is finally complete, a perfect snare and work of art all in one – we are talking superb form and function here.

5: Depending on the species, the maker of the web will lie in wait, either in the centre of the web or secreted off to one side in a shelter or curled leaf. The spider keeps a foot or two on one of the radial silk spokes to detect the slightest vibration created by prey getting tangled.

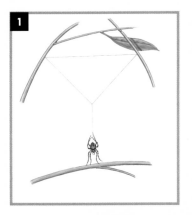

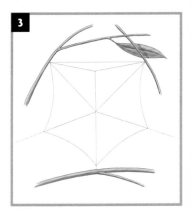
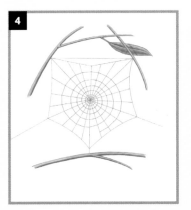
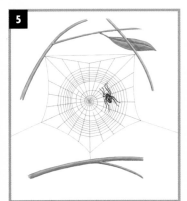

VARIATIONS ON A THEME: DIFFERENT TYPES OF WEB

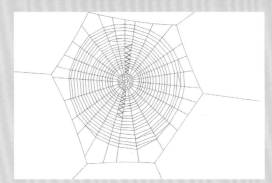

This is a very distinctive take on the orb web theme, belonging to two species in the UK, the Wasp Spider, *Argiope bruennichi* and *Cyclosa conica*. Both spin webs with thick, silk, zigzag stitching worked into the centre. The stitching is known as a stabilimentum and nobody can really agree what it is for.

Many species line their burrows with a silk sock. *Amourobius* spiders make their distinctive funnels in walls, crevices and among the litter. Prey animals get their feet caught up in the matted web and remain stuck there.

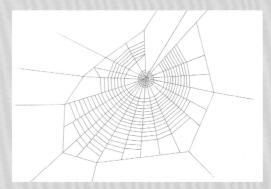

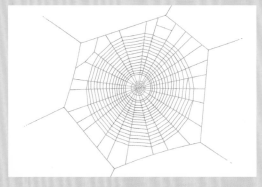

The webs of spiders that belong to the group *Zygiella* look like they started well but the spinner did not bother to finish off one corner. Follow the strand that runs through this 'window' and it will lead you to the home owner, waiting for a bite.

This is the design classic, the orb web of the Garden Cross Spider, the perfect aerial snag, most obvious towards late summer and autumn when these spiders mature.

Experiment: Making a 'tegenarium'

If you want to learn about everyday spider life there is no better teacher than the House Spider. Of all British spiders this is one of the biggest and it is easy to find, study and keep as a pet. A simple tank with a few twigs and some soil and a few flies bunged in regularly is all you need.

YOU WILL NEED:

plastic aquarium with lid • bark, twigs and soil

1: Put your arachnid in the plastic aquarium. These come complete with a lid, built in ventilation and a hatch, which makes feeding your spider easier. Put in a piece of interesting bark, soil and a few twigs. The spider will provide the rest of its abode.

2: A few flies or other small creepy-crawlies every week should provide your spider's food and moisture, but if the place you are keeping your spider is warm then drip a little water in too. Now watch him spin a web before your eyes.

Keep a record of what your spider does, how often it eats and sheds its skin and you will soon get to know these amazing little house guests.

Experiment: Anyone for tennis?

YOU WILL NEED:

• a coat hanger or bendable wire

Spider silk is the strongest natural material and is weight for weight, length for length stronger than steel. But those strands are so thin it's hard to demonstrate this strength. However, if you collect enough webs you can get some idea of spider silk's remarkable properties.

Take a small loop of stiff wire, bend it around so it has a 'spoon'-like shape (a ring with a handle) and on one of those dewy, cold autumn mornings, try collecting a few webs by placing the wire loop behind the web and drawing it forward. The web should stick. Repeat this several times – try to choose those without residents! Soon you will have lined your 'spoon' with pure spider silk.

You can place weights on the collected silk and you will be surprised at how much weight even just a few webs can hold. You can even take a little ball of tin foil and bounce it on the silk you have collected – it shows how strong and stretchy spider silk is. If you have enough silk, you could play a small game of tennis with a friend!

YOU WILL NEED:

spray paint • newspaper • artist's fixative or hair spray • coloured card • scissors

It sometimes seems a shame that those gorgeous orb webs that you find draped in the shrubbery or hanging in the hedgerow rarely last longer than a day. Being fragile and ephemeral is of course part of their attraction and from the spider's point of view they are disposable and recyclable insect traps. However, it is possible to preserve these structures and even hang them on your own wall.

First of all, choose a still day and find a real beauty of a web. Make sure it is dry (no droplets of dew) and make sure its maker the spider isn't in residence (check well in and around the edges of the web especially in curled up leaves) as it really won't appreciate what comes next!

Take a can of spray paint – white or black is a good choice – and holding a sheet of newspaper behind the web to stop you getting paint all over the plants or bushes, spray the web evenly and lightly on both sides from a distance of about 40 cm. Get too close and you will blow a hole in your web. Leave it to dry for a while and repeat.

The next step is to make your web super sticky. You do this with artist's fixative (available from art and crafts shops). This comes in spray cans like paint and in the same way you coloured your web with paint, spray both sides of the web. You can also use hair spray.

Before it dries, take a bit of card, big enough for your web to fit on and of a colour that contrasts with the colour you sprayed your web. This is the trickiest part of the whole operation; you need to line the card up perfectly with the web and push the card onto the silk so that it sticks in the right place first time. Once the web has touched the card you cannot change your mind

without ending up in a messy tangle!

If you've done it right, you should have a perfect web on the card. Use scissors to cut the supporting strands and you can give it another coat of fixative to make sure it's well held in place. You can now mount this spider's original in a frame and hang it on your wall! You could measure all the strands of silk and work out how much silk was needed to make your web and even collect the orb webs made by different species of spider.

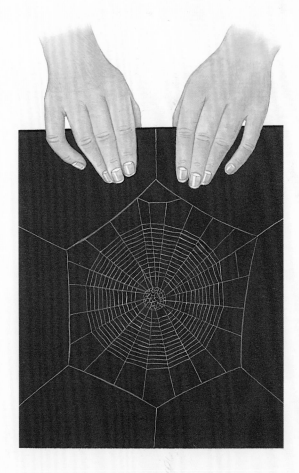

Surfing the web

Believe it or not spiders can be found at sea and in the air – as high as 3,000 m! They form part of a strange world of aerial plankton that drifts around on air currents. Go out on a dewy autumn morning and you can see why their technique is known as 'ballooning'. Just about any long grass will be laced with the dew laden strands of silk, not webs as such but seemingly single strands that link grass blade to grass blade. Investigate closely and you may find the culprits, lots of tiny money spiders, either spiderlings or tiny adults belonging to a family known as Linyphilids. If you collect one of these minuscule animals on the tip of your finger and gently blow on it you may persuade it to 'balloon' for you.

If you are lucky, the spider will raise itself up on tip toes and allow the breeze to pull out a thread of silk, which will snake up on the wind. When this develops enough lift to overcome the weight of the micro-spider, it lets go and drifts off.

Above: By gently blowing on the minuscule arachnid, you may get it to turn its bottom into the wind and lift its abdomen, issuing forth a silk strand. As soon as the resistance of this silk is enough, the spider will sail off.

Harvestmen – the spiders that aren't

Lift ivy that has grown against walls, look amongst ferns and thick vegetation, in dark corners of the shed and under windowsills and you are sure to turn up a gangly life form that when disturbed wobbles off in a manic random sprint that is seemingly better coordinated than its cotton fine legs should allow!

The Brazilian name for them, which I rather like, is *giro mundo*, which refers to their speed and agility. We know them as harvestmen, owing to their apparent arrival towards the end of summer. This is when they mature and are largest and hence more evident.

The technical name for the harvestman family is the Opiliones. We have something like 23 species in Britain but even after a leg count has revealed eight legs, it is a surprise to some to find that they are not spiders at all, although they are arachnids.

They differ from spiders, which they superficially resemble, by not having any poison glands, not being able to produce silk and by having their body parts fused together to form a single button-like body, suspended usually when active between the long vegetation-spanning legs. If you have any doubts about identification, look at the body: a harvestman is an all-in-one affair, no two halves here!

Harvestmen hang around in shady places where it is moist, under window-sills, behind ivy or up against the shed eaves – only emerging to feed by night.

OTHER ARACHNIDS

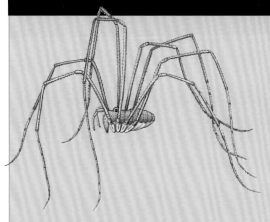

Sheep Tick You are most likely to encounter this parasite in the ears and around the head of your dog after it has been in long undergrowth. These creatures have complicated lifecycles and rely on the chance that a suitable host will pick them up, where they feed on blood. Deer and sheep are the usual hosts.

Harvestman *Leiobunum rotundum* With long lever-like legs and a button-like body, it's easy to see why this common species has many other groovy names, like air crab.

Red Spider Mite The biggest of the mites and one I come across regularly sunning itself on a spring morning on the walls of my garden is the Red Spider Mite, which looks like a small bobble off your granny's favourite curtains that has come to life.

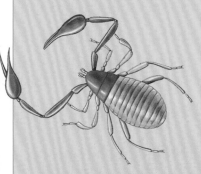

Pseudoscorpion Over 25 species of these little creatures – the biggest is only about 5 mm long – live in the UK. Completely harmless to us, they have no sting in their tail, but they do have pincers loaded with poison glands, to tackle their minute prey deep in the leaf litter and compost. Look for them by spreading out leaf litter on a white sheet and investigating with a magnifying lens.

Harvestman *Nemastoma bimaculatum* Daddy short legs! Not all harvestmen are blessed in the limb department. This one is found among the leaf litter.

Introducing the insects

Insects are the most successful life form on the planet. Scientists think there could be anything from three to ten million species! We know very little about them and it is this that makes even the most basic backyard science exciting. Butterflies, moths, dragonflies, damselflies, crickets, grasshoppers, wasps, bees, ants, bugs, beetles and flies are all insects.

An insect body: the earwig

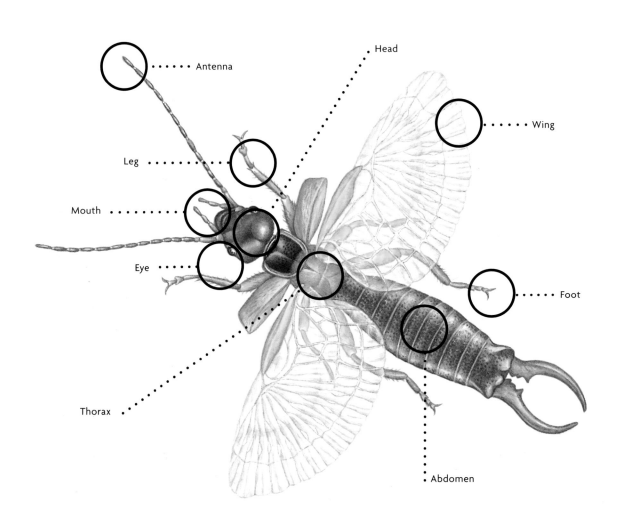

Antenna · · · · · ·

Head

Wing

Leg · · · · · · ·

Mouth · · · · · ·

Eye · · · · ·

Foot

Thorax · ·

Abdomen

What makes an insect an insect?

Despite a confusing array of shapes and sizes, insects all have a basic body plan. All insects have six legs and three main body sections: a head, thorax and abdomen. Of all the invertebrates only insects have wings, but not all insects have wings! If it's a creepy-crawly and it can fly, it is definitely an insect.

Head

The head of an insect is like a box. It is sometimes called the head capsule.

Below: Insects come in a wide variety but all follow the same basic pattern. Wasps belong to the order Hymenoptera, along with bees, ants and sawflies.

Antennae

Sometimes these are called feelers, but they do a lot more than feel and touch: they are also sensitive to smell, taste and vibration.

Eyes

Some insects can detect light through their skin or cuticle but most have specialised eyes. These range from very simple single eyes, called ocelli, that probably just detect whether it is light or dark, to very complicated compound eyes, like a dragonfly's. Compound eyes can have thousands of lenses and are very sensitive to movement and probably detail. Some insects have very good colour vision and can even see ultra-violet.

Mouth

Not all insects have a working mouth. Some moths don't bother feeding – all their time is taken up by breeding! But insects that do feed have a variety of mouth shapes and structures. Some are a bit like our jaws but work from side to side rather than up and down. Some are flattened to crush seeds. Some are like spades or trowels. Some even get stretched to crazy proportions and are used as weapons.

Thorax

This is the bit of the body with all the legs and wings sticking out of it – inside are the muscles.

Legs

Many insect legs arespecialised for jumping, swimming, digging or fighting and many have sensory hairs that can feel, taste, smell and hear.

Feet

Insect feet have lots of spikes, bristles and hooks. Some keep the antennae, eyes and body clean. Others are like grappling hooks, and some have a special pad covered in thousands of tiny hairs that can stick to surfaces.

Wings

Insects that have wings have two pairs, sometimes joined by little hooks so that although there are four wings they act as two. Some insects use their wings for display or to make sounds.

Abdomen

The heart and the organs for digestion and storage of food are found here.

Spiracles

Spiracles are tiny air holes on the side of the body leading to tubes called trachea. These act like ventilation ducts.

Top right: This Pond Skater or Water Strider is a very common aquatic bug. It skims the surface of ponds and lakes but its feet never get wet.

Bottom right: Beetles are the largest order of insects, Coleoptera. There are more than 350,000 different beetle species known to science.

Butterflies and moths

Butterflies and moths are perfect public relations agents for the insect world. The popular image of butterflies as harmless, flouncy, care-free insects flitting idly from flower to flower is helped by the fact that they mainly fly on the sort of days we like – calm sunny ones! The appeal of moths lies in their mastery of camouflage and the mystique of the night.

These delicate animals are subject to the same pressures of living, feeding and breeding as the uglier insects andinvertebrates. They fight their own battles with weapons, nasty chemicals or itchy hairs. They deceive, murder and bribe their way through their lives.

Butterfly or moth?

I can usually tell if an adult or caterpillar is a butterfly or a moth. But if someone asks 'Why?' I start having a bit of trouble. It's a subtle combination of look, movement and shape. So what makes a butterfly a butterfly, or a moth a moth?

Scales on wings

Both have wings covered in tiny scales. These give them their patterns and coloration.

Wing position

Butterflies rest with their wings closed above their body and moths rest with theirs horizontally.

Colours and patterns

Many people think of butterflies as colourful, moths dull and dreary. Find a picture of a Dingy Skipper (a butterfly) or go and look at the pinks and greens of an Elephant Hawkmoth, and tell me that holds true! Generally speaking, though, you do need to look a bit closer for the beauty of moths, while butterflies are much more in your face.

Antennae

The technical name for butterflies, Rhopalocera, means 'clubbed antennae'. Moths are Heterocera, which means 'different antennae'.

Left: Butterflies and moths are popular insects that many people identify with. Some of them are very friendly!

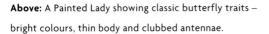

Above: A Painted Lady showing classic butterfly traits –
bright colours, thin body and clubbed antennae.

Above: A Herald Moth. Chunky body, more dowdy and
string-like antennae.

Body size
Butterflies and moths havehair-like scales. These
tend to be a little longer in night-flying moths
and can make the body look bigger. But some
moths have thin bodies, especially the smaller
geometrids and grass moths.

Flying time
Most moths fly by night andbutterflies fly
by day, but there are a few day-flying moths. In
the tropics there are also some night-flying
butterflies.

A word on nets and boxes
I don't really like butterfly nets. It's too easy to
get obsessed with the thrill of the chase and the
final capture, a mentality best left with the
Victorian collectors. I think eyeballs on their own
are best for learning about butterflies and
moths.

But nets do have their uses, particularly when
you are having trouble identifying that little
brown butterfly that keeps whizzing around and
then shooting off like a miniature jet fighter.

With the correct use of a butterfly net and
the right kind of observation pots, you can catch,
bag and identify your beast without even laying a
finger on it.

Try to avoid swiping with the net as this can
damage the fragile wings. Use the net very
gently, picking the insects from vegetation if
possible.

If the insect is flying, swing the net from
behind and, with a quick flick of the wrist, fold
the net to trap the butterfly or moth in the bag.
Then you can lift up the end of the bag and allow
the insect to fly or crawl up toward the light.

Nets, even black ones that allow better views
of your subject, may not give you a clear enough
view of the details you need to see. I tend to

transfer the insect gently to a special cardboard pot with a clear bottom called a pill box. These can be bought cheaply from special entomological suppliers.

Alternatively, you can make your own. Cut up some toilet rolls and tape a cardboard lid to one end. Then make a removable lid out of some clear plastic for the other end. Tape the lid into position so that your specimens don't escape.

Butterflies and moths can be transferred to a pill box by gently cupping it over the insect as it rests in the net. The advantage of the clear bottom is that the insects will crawl toward the light, making it easier to avoid trapping wings and legs when you slide the lid on.

Use a small torch for extra light if you need to pick out tiny details.

Below: Although I'm not a big fan of nets, they can be useful if you need to get a close look at small insects. Always remember to be very gentle.

Metamorphosis or magic?

Butterflies and moths transform several times throughout their lifetime, so much so that they look and behave like very different animals.

Incredible eggs

Butterfly and moth eggs are amazing. Wacky colours and markings abound and some come in such bizarre shapes they look like spiky golf balls or super ornate crystal green houses.

Some eggs are laid singly, others in clumps and clusters, some are sprinkled far and wide and some are plastered with hairs. It is impossible to generalise and show you how to find all of them —but see the box overleaf for some tips.

Crazy caterpillars

Caterpillars come in as many fascinating shapes, sizes and colours as the adult insects. They need to eat and grow as quickly as possible —and avoid being eaten themselves!

A REAL EASTER EGG HUNT

The best way to find eggs is to watch the adult insects, so it helps if you know the food plants of the insects you are watching. Common species such as Small Tortoiseshell and Peacock make egg laying rather obvious. They choose stinging nettles as their food plant and one of the best times to watch out for them is in spring, when the adults have been out of hibernation for a couple of weeks. Look in sunny corners of fields and wasteland where the nettles look fresh and succulent.

When female butterflies are searching for a place to lay their eggs they shop around, keeping low, and touching down briefly to taste the leaves with their feet. When they find the perfect nibble for their caterpillars, they curl their abdomen under the leaf and often keep fluttering their wings. Don't get too close or you will disturb the butterfly, but as soon as she flies away memorise her position and go and look. Gently turning over the leaves on the plant she was on should reveal a cluster of glassy green eggs.

Some butterflies never settle to lay eggs; they scatter their eggs like salt while flying over clumps of their caterpillar food plant. Look up some background information on the species you are interested in to help you decide which plants to check out.

Night time is the right time

Nearly every animal that eats insects is looking for caterpillars, so many have developed neat tricks to help them play the ultimate game of hide and seek, and many feed only at night. The best way of finding them is to invest in a good focusable torch and a set of rechargeable batteries.

Many caterpillars are camouflaged, with patterns that break up their outline and colours that blend into their surround-ings, so it is difficult to see them by day. But at night they relax and move around, often heading for more succulent leaves at the ends of twigs and branches. Using a torch, you can pick them out rather easily.

The secretive and rarely seen caterpillars of the very common brown butterflies, such as Meadow Brown and Gatekeeper, can also be seen at night. You will need to work low down in long grass. Get right down on the level of the tender grass shoots and with a bit of luck you will pick out the shapes of small brown and green caterpillars.

During the day look for signs of chomping – big chunks and crescent shapes taken out of the edges of leaves are a good clue. In cities the presence of real monsters like the caterpillars of Lime, Privet and Poplar Hawk Moths is often given away by the distinctive round droppings scattered on the pavements under suitable food trees or overhanging hedges.

Right: Every year I get phone calls in the summer from distraught people having found a snake in among their fuchsias! It is usually not so at all but the large and spectacular caterpillar of the Elephant Hawk Moth.

Eating for a living

Caterpillars are stomachs with legs. They are completely surrounded by food from the moment they are born and their job is simple: eat all the raw materials that will be needed as an adult moth or butterfly.

Use a hand lens to look at the biggest caterpillar you can find. You may find a cluster of simple eyes (which recognise light and dark only) and near them, either side of the mouth, a pair of tiny downward-pointing antennae to taste and feel the food. The mouth is a marvel of scissor-like plates that mash, mangle and masticate plant matter all day long.

Splendiferous spiracles

Spiracles are tiny breathing holes that all insects have spaced out down the side of their body. They can be tricky to see in other insects, but in smooth, non-hairy caterpillars they stand out a treat.

DON'T BEAT AROUND THE BUSH, BEAT THE BUSH!

One of the best ways of finding lots of different caterpillars (and many other groovy insects) is a technique used by many professional bug hunters, known as 'beating'. The aim is to knock insects out of the bushes or trees in which they live.

You need something to beat the bush with – try a stout stick or broom handle – lots of collecting pots and something to collect the caterpillars on when they fall. You'll be surprised at how many interesting bugs will fall your way.

You can buy beating trays, but an upturned umbrella or an old white sheet will do just as well. The technique is simple: hold your umbrella under the bush or branch of your choice, or spread out your sheet, then take the stick and strike or shake the foliage. If you are too gentle, insects will grip on even tighter and it will be hard for you to dislodge them by further shaking and beating. So make the first strike count!

Left: Moths are not always dull, grey and brown. The Elephant Hawk Moth's bright colours are startling to say the least. They act as a deterrent to inquisitive predators.

How many legs?

A golden rule of insects is that they have six legs, but caterpillars have cheated and developed a few more! They have six true legs clustered near the head, made of hard material and with joints. Another eight legs, called pro-legs, are found toward the back end of the caterpillar. They are soft and end in a collection of Velcro-like hooks.

Showing a softer side

The rest of the caterpillar's body is a flexible and expandable sac, a good example of a moveable and soft exoskeleton.

Most insects have atough exoskeleton supporting their body and providing its structure. In caterpillars the support comes from body fluid, which is kept at pressure. This allows them to have a simple skin which is easy to get out of, meaning caterpillars can grow very quickly by shedding their skins once or even twice a week. The price caterpillars must pay for

Left: Moths are not always dull, grey and brown. The Elephant Hawk Moth's bright colours are startling to say the least. They act as a deterrent to inquisitive predators.

this convenience is they do not have the armoured protection of other insects.

Flags, fuzz, bristles and spines

Caterpillars have an armoury of tricks and gadgets that are spectacularly good at helping them make it to the next stage of life. Look at any caterpillar you come across when beating bushes. The only reason you can spot it is because you have knocked it onto your sheet – put one back on a twiggy branch and it vanishes instantly. These animals do perfect stick impressions, as well as good version of leaves, bark and even, in the case of the Comma butterfly, a bird's dropping!

Above: All those greens and the slanty lines down its side help to break up the outline of this baby Eyed Hawk Moth, providing camouflage amongst the leaves.

Defence

Some caterpillars steal nasty chemicals from the plants on which they feed, giving them a bad taste. Many advertise their unpleasantness with bright colours. Other animals learn from experience that these creatures taste horrible and next time they see one they'll leave it alone.

Some caterpillars have poisonous itchy hairs (which work very well on humans too and will leave you with a painful rash, or very sore and swollen eyes if you rub them after touching one).

Next time you see a group of Peacock or Small Tortoiseshell caterpillars feeding, touch the leaf they are on and watch what happens. These guys have many levels of defence. Their bodies are covered in spines and during their early days they feed as a group, which gives them safety in numbers.

If they are disturbed they will all twitch their heads together, giving the impression of an animal bigger than a mere caterpillar. If this doesn't work they will vomit gobs of green liquid in the direction of their attacker. If these tactics fail, the caterpillars simple roll off the leaf and fall down to the base of the nettles on which they feed, where they are extremely tricky to find.

Right: Just a selection of weird and wonderful caterpillars. I challenge anyone to not be a little scared of those bristles.

Now for something completely different!

If the caterpillar survives for long enough there will come a day when it prepares to moult again. There may be a few clues that this is not to be like any of the skin changes made so far.

Some caterpillars change colour and their brightness disappears. Certain moths will bury themselves in soil, others will weave a silken sleeping bag around their bodies. Butterflies may hang upside down, curling their head up in a J-shape. Some, like Cabbage Whites, remain head up, but spin a little waist harness around their middle.

Garden Tiger

Pale Tussock

Peacock

There is nothing more perfect than the pupa of a moth. All conker shiny, red, brown or orange, many sit out the winter months in this stage of their life cycle between caterpillar and adult moth. To find some take an old margarine tub of soil or moss, a trowel and a hand fork, and gently dig around the bases of trees. Do not tear up the soil or go any deeper than 10 cm.

Be patient and you will find the gems you are after. In my experience the best trees are the willows, sallows and poplar.

Sterilise some soil or moss (10 minutes in a microwave oven), place it in a seed tray or tub and arrange your pupae collection on this. Keep it in a cool place such as a shed or garage out of the sun and spray with water once a week, making sure that the soil is always damp but not soggy. Place a few sticks in the soil so the moths have something to climb up.

If you have found your pupae in the autumn or winter, they will probably not hatch until the following spring, but any other time of the year and you will just have to keep checking, they could hatch at any time.

Then the head capsule pops off, and the skin splits and is shrugged off, revealing something that looks like nothing alive. It has no features to speak of, although if you look close enough you can make out in some species the strange mask-like impressions of legs, eyes, antennae and wings of the animal to be. If it wasn't for the odd twitch and wriggle, there is nothing that would give you a clue that this is a life form.

This is stage three in the lifecycle of a butterfly or moth. What has just been created is the pupa or chrysalis. Inside it one of the most mind-blowing transformations in the world is happening as the caterpillar becomes an adult butterfly or moth. It dissolves into a rather disgusting living soup, a gunk of cells, and rebuilds itself as another animal altogether.

Although this event is worthy of a fanfare, it all occurs very quietly – a chrysalis or pupa is a tasty snack for many animals and so needs to stay hidden. Some have amazing sculptural qualities, particularly the exposed chrysalises of most butterflies, with weird lumps and bumps that break up the outline and make them difficult to see.

Others are camouflaged and some even play tricks with mirrors. The chrysalises of many butterflies have shiny panels that reflect the background and bounce light around. Some are disguised as leaves.

Moths go underground or surround themselves with a fortified cocoon of silk, produced by glands in the caterpillar's mouth.

PUPA OR CHRYSALIS

Pupa means 'doll or puppet' and can be used for both moths and butterflies. Chrysalis means 'golden', referring to the metallic sheen on some, and is used exclusively for butterflies.

Comma lifecycle

The eggs of the Comma butterfly, almost ready to hatch.

The caterpillar feeds constantly to fatten itself up for its metamorphosis into a butterfly.

The chrysalis is camouflaged as a dead leaf to fool predators.

The pupa emerges as an adult butterfly and will soon start taking nectar from flowers.

Flamboyant finale: the adult

Finally the pupa splits, revealing a creature nothing like the caterpillar. It has a weird head, consisting entirely of eyes – big ones too, not the simple specks of the caterpillar but large and googly.

There are long antennae and, in most species, a tube-like mouth that is coiled up like a hose pipe and held under the head. This mouth starts of as two halves of a tube, which join together soon after the adult emerges.

If you actually get to watch this event, just take a moment to look at the new adult. Without wings, which on emergence are all scrunched up, they really are quite ugly. At first this creature is trapped between two worlds. But wait a while longer...

With six spindly legs it drags itself up to a position where it can hang freely. It now pumps blood into the veins in the wings, which expand. The liquid hardens and sets, and then a pair of stiff wings is ready to fly!

Tortoiseshell lifecycle

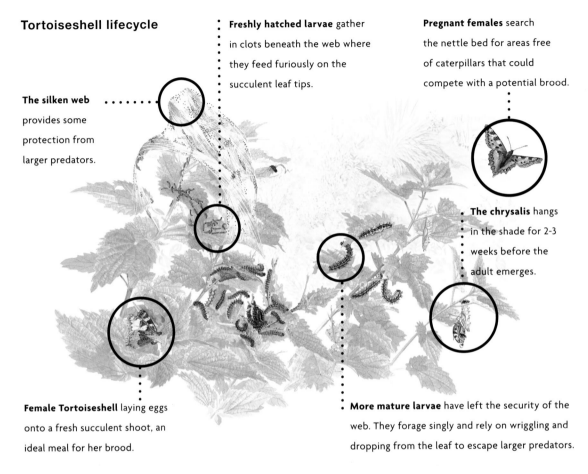

Freshly hatched larvae gather in clots beneath the web where they feed furiously on the succulent leaf tips.

Pregnant females search the nettle bed for areas free of caterpillars that could compete with a potential brood.

The silken web provides some protection from larger predators.

The chrysalis hangs in the shade for 2-3 weeks before the adult emerges.

Female Tortoiseshell laying eggs onto a fresh succulent shoot, an ideal meal for her brood.

More mature larvae have left the security of the web. They forage singly and rely on wriggling and dropping from the leaf to escape larger predators.

Wings – not just for flying

A butterfly's wings are survival devices and do much more than propel the insect through the air...

Solar panels

Wings are used like radiators. Butterflies need to get their body temperature up to around 30°C to be active. Held open, wings can be used as solar panels, to warm the blood and take it back to the insect's body. This is the familiar basking pose adopted by many butterflies when the air temperature is cool, during the early morning hours.

Team colours

The bright patterns of the day-flying butterflies and moths act as signals to others of the same species, which is particularly useful when trying to find a mate.

The colours and patterns of most moths help them hide during the daylight hours. Bright wing colours also act as a warning. In most species sombre colours on the other side of the wing can hide these flash markings.

Attack deflectors

Eye-spots, such as those seen in the Peacock Butterfly, can deflect an attack. If you study a population of Meadow Browns or Gatekeepers you may notice that a lot of them have triangular bits of their wings missing. These are places where birds have aimed at spots on the wings, and a just a bit of wing has broken away allowing the butterfly to fly another day.

This is taken to the extreme in the Blue and Hairstreak butterflies. The undersides of their wings also have an eyespot at the corner of the rearwing, and some species have another blotch of colour and a tail, which looks just like a second head with antennae. Predators lunge for the dummy head instead of the real one – the butterfly could end up losing a bit of its wing, but at least it has not lost its head!

Perfume wafters

A surprising use of wings is as wafters of sexy perfume. Many male butterflies have scent scales that break off from the surface of the wings. These are filled with a perfume used in courtship. Some male butterflies draw the female's smelling organs, her antennae, over the scales. Others simply vibrate their wings at the female.

A distinguishing feature of male Gatekeepers is a dark smudge in the middle of the forewing. This is a patch of those very special scales.

Left: The bright forewings of this butterfly are like shiny copper, hence its name Small Copper. Its caterpillars are small and green. It flies during summer months.

Experiment: All sweetness...

Moths are notorious boozers, so a great way to see some species is to provide the drinks. Lepidopterists have used this technique known as 'sugaring' to attract moths for hundreds of years.

You need to make a cocktail, and it doesn't matter all that much what you put in it as long as it is sticky, sweet and smelly. I have had great success with this recipe..

YOU WILL NEED:
big pan • sugar • molasses • fruit juice • dark rum •
water • paint brush • jam jar • torch • rags

1: Mix together all the ingredients.

2: Gently heat and stir adding water or sugar to get the consistency quite runny.

3: All you need to do now is to wait until dusk, preferably on a warm summer's night. Take a jam jar full of the sugaring solution and a stout paint brush and apply the mix generously to tree trunks and fence posts in your garden. If you vary the positions and habitats, you are more likely to see more kinds of moth.

4: Return to your posts an hour or so later with a torch and see what you have lured out of the night.

5: An alternative is to make a runnier mixture and soak old rags in it. You can hang these up around the place to attract moths.

... and light!

The other well-known attraction for moths is light. Nobody really knows why they fly towards the light but the most convincing idea is that nocturnal insects use light from the moon for navigation when flying at night.

Try it for yourself. On a moonlit night walk with the moon in the same place in your field of vision and you will move in a straight line. Then get a friend to stand in one spot with a torch. Try and keep it in the same place in your field of vision while you walk and you will find thatyou will need to turn your body as you walk.

Eventually you will end up bumping into your mate. This is because the torch is much closer to you than the moon. When humans started lighting up the night with lamps, candles and bulbs, the moths got confused!

This situation is exploited by moth-lovers to find out which moths are around. They use a special light positioned over a trap with a collection bucket and a funnel of some kind. You can buy traps from entomological suppliers but they tend to be expensive.

Above: Most moths and butterflies have a very sweet diet to give them the energy for all that flying. Just give them some sugar and they'll come to investigate.

Become a butterfly farmer

One way of really getting to know these bizarre beasts and the secret chapters of their lives is to rear them through the stages to the adult insects. I think that everyone should at least have witnessed the emergence of a butterfly from its chrysalis before they die – and with the right species it is easy.

To get going you need a few basic bits of kit and the animals themselves. You can't just go into your local pet shop and pick up half a dozen caterpillars, but there are specialised breeders all around the country and if that fails then you can collect your own using some of the tips given in this book.

Above: Garden Tiger Moth. These moths are rarely seen, although they are common and widespread. The best way get a look at one is to try and lure one to a light.

First, make your choice

With 60 regularly occurring British butterflies and 800 or so large moths, there are a lot to choose from! But there are a few that are ideal to start with. These will provide spectacle, not be too fussy or have a common food plant. What is more, they are likely to succeed and not cause disappointment.

Food plant availability is first priority so choose carefully. I also think it's nice to rear a species that you have found yourself in your own patch. This means you have the right food plant nearby and that if you are a little bit successful with keeping them and end up with a shed or bedroom bursting with too many beautiful creatures, then you can easily release a few back into the wild. (By the way, many people think that by rearing a species and releasing it into the wild they are helping – they are not. Trust me on this one, releasing a few will not hurt and the insects will live a natural life, but it isn't going to help the species in the long run.)

Fresh eggs

Once you have obtained the eggs of either a butterfly or moth, keep them in a small airtight plastic box to protect them from predators, and to stop them drying out. You do not need to make air holes, there is enough for many days in the box, but it is a good idea to open the box to ventilate and breathe on the eggs gently (to keep them moist) every other day.Then simply wait and watch for a clue to the caterpillars' hatching. They often do this at night, so check every morning. Keep the box at room temperature and do not leave it in sunlight.

Hatch and go

One morning you will be rewarded with the sight of tiny caterpillars only a couple of millimetres long. These little chaps shouldn't be rushed off their eggs, as some need a first meal of egg shell to start them off. Wait until they have wandered away from the eggs in search of food, then use a fine camel hair brush and a spoon to gently transfer them to a slightly bigger plastic box, lined with tissue paper and with a fresh leaf or two of the food plant (do not choose the newest shoots, which will give them a bad stomach, or the tough leathery leaves, which are too tough to chew).

Don't put the caterpillars in a large container as they might wander off their food, get lost, get hungry and die. The perfect containers are very small margarine tubs, but make sure that they

have been well washed out. At this stage do not worry about air holes, the caterpillars will only escape through them!

Keep checking the eggs until all have hatched. From now on the priority is to keep the food fresh and plentiful. Scrupulous daily cleaning is essential. All is well when the caterpillars' dusty little droppings known as frass appear on the tissue paper.

All change

After a couple of days, the caterpillars will shed their skins for the first time. When waiting to moult the caterpillars will appear very still. When they get bigger there is also a bulge just behind the head capsule. Do not disturb them, simply place the old leaves on some fresh ones. The caterpillars that are hungry will move; those moulting will do so when they are ready. They often change colour at this stage and spin themselves a little pad of silk to which they attach their back legs. This helps them to walk out of their old skin – the caterpillar equivalent of standing on the leg of your jeans while you pull the other leg out! Once they have crawled out of their old skin they will resume feeding.

Growing up

After a week or so you will probably want to re-house your beasts, as they will be growing fast and producing more and more frass. They will start to produce more moisture themselves, which means it is time to increase ventilation. The next suitable house can be a bigger margarine pot, or a stout cardboard box such as a shoe box.

MAKING YOUR CATERPILLAR BOX

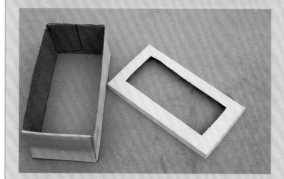

Simply cut a hole in the lid, so the edges make a frame. Then glue or staple a fine mesh, gauze or muslin over this, making sure the holes are not big enough to let the caterpillars through!

Cut stems of the food plant can be placed in a small bottle or jar of water. Film canisters or spice jars are conveniently small and it is fairly easy to cut a hole in the lid for the plant stems. To prevent the caterpillars from drowningstuff tissue paper or plasticine in the gap around the stem.

As they grow, continue to clean and refresh the food on a daily basis. If necessary upgrade to a bigger box.

Re-mould and reshape

After five or so skin changes your caterpillars should be nearly grown up. Just prior to moulting for the last time into the pupa, many species get an uncontrollable wanderlust and go charging around the cage paying no attention to feeding. The often change colour at this stage. You must make sure you provide the right conditions for them to form a chrysalis or pupa.

Most butterflies need sticks or twigs on which they can dangle or which they can brace themselves against. But moths, depending on species, often need either some egg boxes or a layer of soil in the bottom of their cage. The general rule is hairy caterpillars need egg boxes and smooth ones need soil, but there are exceptions to this.

If you are using soil make sure it is sterilised, I usually give it a blast for 10 minutes in the microwave. This kills any fungus, predators or diseases that may be lurking. Let it cool and dampen it down with water again before placing in the bottom of your rearing cage. Big caterpillars will need a layer of soil at least 10–15 cm deep.

Playing the waiting game

The pupa or chrysalis now needs tending. This is the bit where you give an inanimate object love and care and, for now at least, it gives you nothing back.

In the wild this stage can last for anything from a few weeks to months, depending on the species' annual cycle and often the temperature the pupae are kept at.

Your chrysalis or pupae should now be placed

Above: A night flyer, the Leopard Moth occurs in gardens, parks and woodlands.

in an emergence cage. This has to be spacious and well ventilated with branches, twigs or netting for the emerging adult to climb up.

Spray with water every few days, using a plant spray and keep the pupae out of the way of predators, especially when storing for the winter.

A star is born

Eventually your prizes will emerge. The pupa or chrysalis darkens in colour, then goes semi-transparent – sometimes so much so you can see the patterns of the wings through its shell. Now is the time to keep checking back every few hours, especially during the early morning. Emerging at this time gives the adult butterfly or moth time to crawl to safety and dry its wings under cover of darkness. Having completed the cycle, the adults can now be released back into the wild, or if you have a male and female then you can keep them to see if they will mate – and you can track the whole cycle again.

A few species to rear at home

As you get more experienced, you can try rearing other species, but I recommend getting a book or two on the subject first for more technical advice.

Poplar Hawk Moth

Lime Hawk Moth

Large White

Small White

Eyed Hawk Moth

Large White caterpillars group together and are coloured bright yellow, white and black as a warning that they are nasty to eat, but Small White caterpillars are green and hang out on their own.

Lime, Eyed and Poplar Hawk Moths are good to start with, especially as the adult moths do not feed, which cuts out one of the most difficult bits of breeding butterflies and moths!

Small tortoiseshell

Red Admiral

Small Tortoiseshells feed on nettles. The best way to collect stock of these is to look for the egg-laying adults in spring. Do not collect the whole batch of eggs, just take 15–20.

Red Admiral caterpillars can be found by searching nettles for leaves that have been folded into a tent (not rolled – these usually belong to the caterpillar of the Mother-of-pearl Moth).

Dragonflies and damselflies

Dragonflies were once called 'horse stingers' or 'devil's darning needles'. These are great names but don't be fooled and think that these insects are nasty pieces of work that sting – they don't!

Dragonflies' extrovert tendencies, the fact that there are relatively few species in this country (38 species regularly breeding in Great Britain and Ireland) and their relatively large size, make them fairly easy to get to grips with. The catch is that most are fairly fast fliers – larger dragonflies can reach speeds of 36 kph (even damselflies can reach 10 kph), which makes following them with the eye next to impossible.

The hardware

The scientific name for dragonflies is Odonata meaning 'toothed jaws'. These insects arethe winged assassins of all other small insects and even the tiniest of damselflies is a ruthless murderer of midges – some even consume up to 20% of their own body weight a day.

As nymphs (larvae) they are masters of ambush, lying in wait in the mud and leaves on the bottom of ponds or amongst the water weeds. Their secret weapon is one of the most complex hunting gadgets – a set of extendible jaws or, strictly speaking, a lethal lower lip.

This deadly device is hinged under the head and when the nymph sees potential dinner (a passing tadpole, fish or another nymph), it extends its lip and stabs, grabs or spears the prey so fast that the action cannot be followed with the naked eye.

But as evil as these little aquatic aliens may appear, it is the jaws of the adult which gave rise to the family name Odonata.

Distinguishing dragons and damsels

The dragonfly family is unmistakable: four large wings, large eyes, and a long colourful body. It's telling apart the two groups – damselflies and dragonflies – that can be tricky.

The easiest way to distinguish between them is to imagine a knight in shining armour turning up to rescue a fair maiden from a beastly dragon. The dragonflies, like their namesake in legend, are bigand bold with an aggressive direct flight. Damselflies are smaller, more gentle, and their floppy flight is reminiscent of the fluttering handkerchief of the damsel in distress.

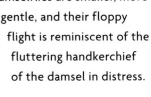

Left: Do not expect all dragonflies to stay in one spot as long as this one!

Dragonfly bodies

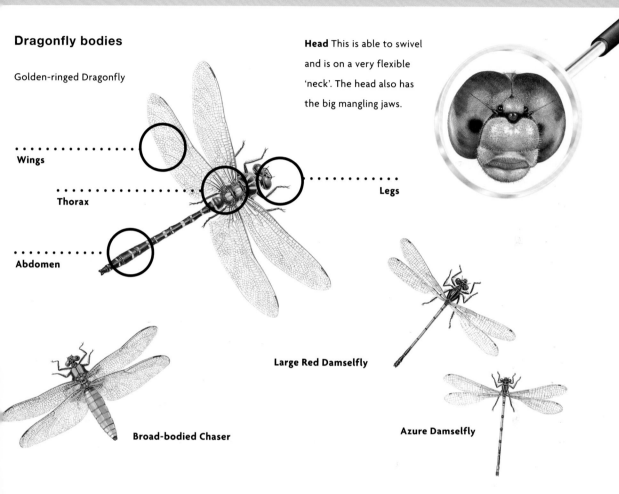

Golden-ringed Dragonfly

Wings

Thorax

Abdomen

Head This is able to swivel and is on a very flexible 'neck'. The head also has the big mangling jaws.

Legs

Large Red Damselfly

Broad-bodied Chaser

Azure Damselfly

Of course, there are lots of technical differences between the two but I always find that for beginners that little story helps identify the overall difference.

Incidentally, the name dragonflies is used by scientists to describe both groups so if in doubt call 'em dragonflies and you'll always be right!

Size

As a general rule, dragonflies are bigger than damselflies. This rule works well in Great Britain, although in other parts of the world this isn't always the case.

Wings

Two pairs of wings can be used independently of each other, giving them the agility and appearance of a helicopter, enabling them to hover, achieve vertical take off and landing and fly backwards. At rest, dragonflies hold their wings flat, while damselflies often rest with their wings folded up over their backs.

Eyes

Dragonflies have the biggest eyes of any insect, big boggly compound eyes that take up most of the space on their head. They are made up of up to 28,000 tiny simple eyes all clustered together. Both insects have very good colour vision. They need to be able to see well to catch insects in flight. The dark bit in the middle of the eye, called a pseudopupil, is the area that can see the best.

Antennae

Because they spend most of their time flying and are rarely in contact with a hard surface, dragonflies and damselflies do not need feelers to feel. Instead they are very visual creatures with antennae reduced to what look like tiny car radio aerials.

Thorax

Dragonflies experience incredible forces through their wings when in flight, so they need big and powerful muscles to work them and row through the air, hence a big engine box of a thorax.

Legs

The six legs all stick forward from the body and each one is covered in big spine-like bristles. This makes them okay for perching and useless for walking, but when held together on the wing they become a death trap, able to scoop up and catch other small insects in flight.

Abdomen

This contains all the organs that the dragonfly needs for digestion and breeding. It is also the most colourful part of the body and the colours and patterns are essential for identification. Females tend to be different colours and patterns from the males and not quite as gaudy.

Be careful in your identification of immature adults as these can often be a different colour from the fully mature adults all together –a good field guide is essential!

Behaviour to look out for...

Egg laying

Different species use different egg-laying techniques. The most obvious is 'stabbing', when the females of darter and emerald dragonflies can be seen 'pogoing' across the water, appearing to be bouncing on the tip of their abdomens. The simple dipping action of the female's abdomen washes the eggs off and they slowly sink below the surface.

Another method of dispersing eggs is carried out inside pond plants, by cutting a small slit in the plant and inserting an individual egg. This is what is happening when you see a female resting on vegetation and probing around with the tip of her abdomen. Some female damselflies even drag themselves completely under water down the plant stem and what's more they often pull their mate too, still attached by his claspers!

Territorial dog fights

Dragonflies and damselflies, in particular the males, can be aggressive and very possessive of a territory. Their only goal is to ensure they mate with and fertilise as many females as possible.

The females wander widely as they mature and feed themselves up, but sooner or later they have to return to the water where the males are waiting. Males find a good perch to survey as much of the water as possible. They will chase and try to impress the girls when they arrive and some of the chases you see will be of this kind.

But other males of the same species have the same plan, which means the sky over the pond or river becomes a battle zone during the height of summer, with males trying to drive other males away in spectacular aerial dog fights, with lots of chasing and clashing of wings.

Mating

This can appear to be rather confusing in most species as they adopt a strange position known as the wheel. Go down to any duck pond in the summer and you will see some damselflies joined together in tandem.

The male, just before mating, pops his sperm from the tip of his abdomen into a special pouch right at the other end of his abdomen on the underside of the second segment. When a pair meet up and like the look of each other, the male grasps the female's neck with a pair of finger-like projections on the tip of his bottom, called claspers. Then she reaches forward with her abdomen to get the sperm stored in the male's special pouch – this results in the wheel

The dragonfly lifecycle

Mating The male grips the female behind her head and she curls her abdomen up.

Larval stage Nymphs spend up to two years under water, where they are voracious predators.

Laying Eggs are laid inside plant stems, on submerged plants, or directly into the water.

Transition The nymph crawls up vegetation and a pale and soft adult with crumpled wings emerges.

Adult stage The fully primed, expanded and coloured up adult insect, ready to start zooming about murdering flies!

position. Impressed? Well, some of the damselflies can maintain this for over six hours, and even fly around in this position!

Feeding

Adult damselflies and dragonflies catch their food on the wing using their special basket-like legs, but occasionally you will find one that has caught a large insect and taken this sizeable meal to a favourite perch to eat it.

Hunting is carried out in a systematic way. Some fly up and down over the water, others can be seen flying backwards and forwards, quartering woodland rides and skimming the tops of low-growing plants. Others rest on their perch and pounce on insects, returning to the perch seconds later to eat and wait again.

A damselfly nymph

Wing buds Look carefully at your nymph. If it is a later moult, the chances are it will have very obvious wing buds on its back.

Size and shape Size is very variable, depending on moulting stage. Shape is a more useful indication. Dragonflies come in different shapes varying from short, fat and flat to long and torpedo-like, but damselflies are nearly always long and almost spidery in appearance.

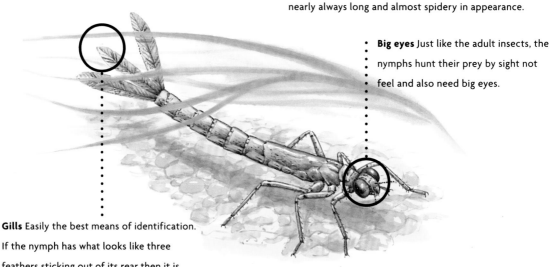

Big eyes Just like the adult insects, the nymphs hunt their prey by sight not feel and also need big eyes.

Gills Easily the best means of identification. If the nymph has what looks like three feathers sticking out of its rear then it is almost certainly a damselfly. These are the gills. Look at the head as well because mayfly nymphs also have three gills, but their head looks different. Dragonflies have gills in a hollow in their abdomen.

Jaws Owners of the meanest gnashers in the pond, both dragonflies and damselflies have complicated mouthparts that they can shoot out and forward to catch their victim by surprise.

Dragonflies and damselflies often perch way out over the water on overhanging trees and other hard to get to places. This little trick brings them to you. It plays on the highly territorial behaviour of males, which like to find a good perch to survey as much of the water as possible. All you need is a stick (a bamboo cane or a pea pole), binoculars and patience. The more sticks you set up the higher your chances of success.

Simply find a body of water with plenty of dragon and damselfly activity. Choose a sunny summer's day and a bank without much tall natural vegetation and stick your stick into the bank or mud – ideally leaning slightly away from the bank and out over the water. This means that any dragonflies that use your perch will always be with their backs to you, making it easier to look at their colours and wings.

Then it is simply a case of sitting, watching ... and waiting!

Experiment: Hatching your own

Set up a fish tank with no lid in a light spot, out of direct sunlight (outside is best so that the insects will be free to fly away). Fill it with clean rain/pond water and add pond weed and small pond life. Add a few branches sticking out for the insects to crawl up.

Find some nymphs in a still-water habitat. You only need 2 or 3 large specimens for a fish tank about 50–60 cm long. Feed your nymphs on small worms. When they are about to change into an adult they hang around near the surface as their gills stop working and they start breathing through spiracles.

Now watch closely in the early morning and evening. If you are lucky you may witness one of the most beautiful sights in the insect world as the nymph splits open along a natural weak spot at the back of its head and a brand new dragonfly oozes out of the husk. If you miss the emergence, try again! (It took me five attempts.) If you do miss it, at least you will be left with the nymph's skin as a souvenir.

Top to bottom The amazing sight of a dragonfly emerging. The nymph crawls out of the water and then splits open so an adult can squeeze out, leaving an empty husk still clinging to the stem.

Dragon hunting – getting started

The avid dragon and damsel watcher needs a few bits of equipment. A pair of close focus binoculars is very handy, because a lot of the action takes place in the middle of the pond or river. A magnifying lens is essential for looking at nymphs and empty skins close up.

Pond nets are useful for reaching into the water to collect and find nymphs, and several clear pots are essential if you want to really appreciate what monsters these young insects are. A white tray for sorting them from the weed, mud and gravel is handy too. A good field guide will help you to put names to everything you find.

Crickets and grasshoppers

Few habitats don't hum, scratch and buzz to the tune of crickets and grasshoppers in summer.Some have an uncanny ability to throw their 'voices' so even when you think you should be getting close, the chirping always seems to be coming from the next clump of grass!

This group contains beastly predators and gentle vegetarians, masters of disguise, and some of the loudest shouters in the countryside, so I'm sure you will want to get to know them better.

Telling them apart

These insects are all very good at hiding, so it would be a lie to say that getting to know them is easy.

As a group they are very diverse: some are small and look like stones, others do great impressions of plants, some fly, others don't. Grasshoppers and crickets are famous for jumping and to perform this act they all have big back legs, to ping themselves through the air.

Scientists argue about exactly what to call them. Some say Orthoptera, which means straight-winged; others say Saltatoria, which comes from the Greek word that means to jump.

Separating the crickets from the grasshoppers is really easy.

Right: A good place to start looking for grasshoppers and crickets is, you guessed it...grass!

Antennae

Crickets have wispy, whip-like antennae, which are longer than the body. Grasshoppers' antennae are very short and much thicker. Crickets are nocturnal and probably need longer antennae to feel their way around and find food.

Ears

Because sound is used in communication and for finding a mate, crickets and grasshoppers have 'ears' called tympanal organs —similar in principle to our ear drum. In crickets they are found on the front pair of legs. In grasshoppers the 'ears' are at the base and sides of the abdomen, but don't bother looking as they are nearly always hidden by the wings and are tricky to see.

Legs

The last pair of legs are larger and contain strong muscles that generate the jumping force. Bush crickets are armed with lots of spines and spikes to strike and pin down smaller prey insects.

Cricket

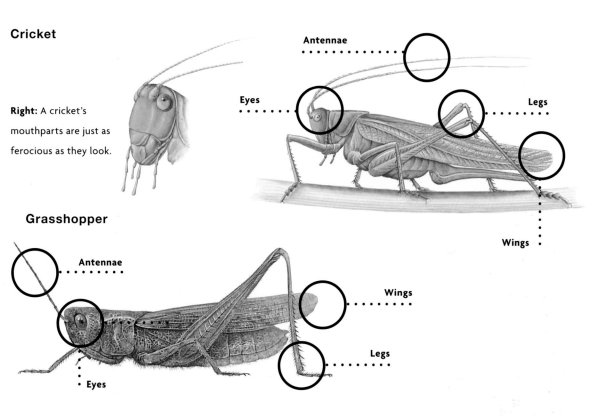

Right: A cricket's mouthparts are just as ferocious as they look.

Antennae

Eyes

Legs

Wings

Grasshopper

Antennae

Wings

Legs

Eyes

Mouthparts

The jaws are like a big set of secateurs or meat shears, and are perfect tools for mashing up both plants and animals. Palps either side are used for feeling and tasting food.

Wings

A frail pair of pleated and folded hind wings are hidden behind and protected by a pair of leathery forewings.

Stridulating organs

Grasshoppers make noise by 'fiddling' using their hindlegs as a bow, rubbing a set of pegs on the inside of the leg against a raised vein on the forewing.

MAKING CRICKET CATCHERS

During the warmer months it is not hard to find the green nymphs of Speckled Bush Crickets and the brown young of Dark Bush Crickets, sitting still in bramble patches. It's not so easy to catch them. One false move, and before you can say 'Blasted blackberries!' the cricket has hopped it.

The answer to your troubles lies in a long pair of scissor-like barbecue tongs. Using strong duct tape, attach a tea strainer to the end of each half of the tongs, so that they close together to form a little mesh cage. With this you should be able to bag crickets, grasshoppers and many other nervous insects.

HOW TO MAKE YOUR PARENTS FEEL OLD...

Most cricket and grasshopper songs are well within the audible range of a human being but as we grow older our ears lose a little of their ability. From their early thirties onwards most adults listening for the song of coneheads, a kind of bush cricket, cannot hear it, but their children will!

Crickets make their noise by rubbing a serrated rib of one wing against the rib of another wing. A clear, roundish area on the wing, called a mirror, amplifies the sound.

The call of the Oak Bush Cricket is made by vibrating one of its rear feet against the surface of the leaf on which it sits. The large Marsh Grasshopper kicks its wings with the spines on its back legs, making a popping sound.

Ovipositor (the bit that lays the eggs)

The egg-laying tube gives crickets their technical name of Ensifera, from the Greek word meaning sword-shaped. You can see this structure in most female crickets, even young ones, and it is a good way of telling male from female. Grasshoppers have smaller and less well defined ovipositors and it's a lot harder to tell male from female.

Grassroots symphony

Why do crickets and grasshoppers sing? Well, living in dense vegetation and hiding from predators makes finding your mate a hard job.

Like small song birds, crickets and grasshoppers sing to attract females and to establish their territories. It is usually the males that make all the noise. The females are generally mute, although some species make a quiet little squeak.

Each species can be identified by its calls: a serenade for the ladies, a war cry when they come across a rival male and a territory song to say 'out of my patch!'

Above: Grasshoppers chirp by rubbing one leg against their forewing, a bit like a violinist using a bow against a string.

Above: Crickets chirp by rubbing pegs on one wing against the other. The sound is amplified by a structure in the wing surface called a mirror.

BUSH CRICKET LIFESTYLE

The lifecycle of crickets and grasshoppers is very similar. It is descibed as incomplete, because it does not have larval or pupal stages. First they over winter as an egg.

After hatching, these insects look like miniature adults, but their sexual apparatus and wings are not fully developed.

They shed their skin a number of times before they reach adulthood.

In most species adulthood is marked by fully formed wings.

Perfect miniatures

Grasshoppers and crickets do not have a four-stage lifecycle, like the butterflies and beetles. They start off as an egg and then go through a series of moults that, from hatching to the final product, are all recognisable as the adult insect.

The clue to the age of the insect comes from the size of the wing buds or pads, which you can see if you look at the insect's back. Only fully grown adults have fully developed wings, normally as long as or longer than the insect's body.

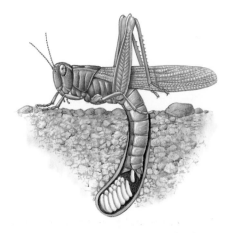

Above: Female grasshoppers lay their clutch of eggs deep in the soil.

One of the best ways to see grasshoppers or crickets maturing is to rear a few nymphs in a plastic aquarium with a fine netting on the top. Don't leave the tank on a sunny window ledge – the sun will soon fry the inhabitants. Instead, mount a 60 watt bulb on the inside of the tank, or shine a desk lamp over it. Use a timer switch set to the same sort of daylight time as is occurring outside.

Grasshoppers can be fed a varied grass diet. Tie a small bundle of mixed grasses together on a piece of string and lower it into the tank. Once the grasshoppers have sussed there is fresh food, they will move to it and the old bunch can be removed.

Crickets need to be fed a very different diet. Give them plenty of variety: bread soaked in honey, grasses, flowers, fish food flakes, fruit, live insects such as young grasshoppers and blowflies (easy to buy from angling shops).

Put in some twiggy branches to give the insects something to cling on to when they moult. The sight of a grasshopper or cricket hanging itself upside down and splitting its neck open and a perfect pale and soft insect gently falling out has to be seen to be believed, especially the moment when the antennae are pulled out with such force that you'll be convinced there will be a twanging sound as one of them breaks!

Depending on the species, they will shed their skin between four and ten times before reaching adult size.

Although the adults may lay eggs in soil provided in your tank, and it is fascinating to see them do this, some species take several years to hatch, which makes keeping them all the way through their life-cycle a bit of a challenge for all but the most dedicated fan!

Finding and catching your quarry

If you sing to advertise your presence it isn't just the intended audience that will hear your call. There are lots of eavesdropping animals out there eager for a crunchy snack, so to avoid being caught crickets and grasshoppers are good at pretending to be plants. Many use a combination of greens, browns and greys to blend into their backgrounds.

Others, such as ground hoppers, do passable impressions of pebbles, while the conehead crickets rest up during the day, drawing their legs and antennae together and tucking in tight to plant leaves. Some simply hang out in the densest vegetation and it can be very frustrating to hear a Great Green Bush Cricket singing its

wings off and simply not being able to see it!

If you do eventually get a sighting it is often followed with a 'ping' as those famous jumping legs are put to good use. To study these animals we need to overcome all these defence strategies. Here are several little tricks for outwitting grasshoppers and crickets. Which trick you use depends on which species you are looking for.

Grasshoppers tend to live in grassy field edges and meadows and you can certainly home in on their song. But you are probably missing out on the females, young nymphs and many other species. These can be swept for using a sweep net, a bit like a butterfly net but with a stronger white cloth bag and a stout frame and handle.

You can buy these or have a go at making one yourself. Sweep the net through long grass a couple of times and then investigate the bag – it normally isn't too long before you have found a grasshopper or two. Sweep different habitats and you will find different species.

Bush crickets tend to be a little trickier because they prefer bushes or at least thicker vegetation. For these you can use the beating technique (see page 65) or you can use a method known as 'walking up'. This is nothing more than walking along like a heron, keeping your eyes in the vegetation in front of your feet. After you have got your eye in you will be surprised at how easy it is. Locate the rough whereabouts of a singing individual and then look for a bit of grassy habitat nearby to walk through.

Another superb way to find them is to go out at night with a torch.

Shed light on a secret

Oak Bush Crickets are active at night so you rarely see them. They don't sing, but do a little tap dance with their rear legs. You can often find them by beating small bushes in the same way you find caterpillars. But another trick relies on their attraction to light. You can lure them down out of the leaves by strapping a powerful torch with long life batteries to the trunk of an oak tree, making sure the beam points up. Come back after it's been dark for a few hours and you could have one or two sitting in the beam.

LIVING THERMOMETERS

Because they are cold blooded, the activity patterns of grasshoppers and crickets are often tied to the temperature of the surrounding environment. A good way of demonstrating this is to listen to them. Just listen to the scratching, chirping and clicking in a field on a sunny summer's day. Go to the same meadow on a dull, overcast summer day and it will be very quiet indeed.

On hot days bush crickets sing an almost continuous trill, but as the air temperature drops individual chirps can be detected. You can even tell the temperature if you count chirps for 15 seconds. To find the air temperature in Celcius (within a couple of degrees) divide this number by 2 and add 6.

Wasps

Do not fall into the mind trap that all wasps are yellow and black beasties out to get you. It simply isn't true. Wasps will, however, graze their way through a colony of greenfly in a couple of hours, a truly disgusting scene, but one that will please many gardeners in its ruthless efficiency!

Sweety or stinger?

I think wasps get a raw deal. A little sting, which is only rarely unleashed on us, is a small price to pay for all their benefits. If they didn't have that sting, we would all love these animals to bits!

Only wasp larvae eat flesh, the adults have a penchant for the sweet and sickly. Remember, one of the things we really like about bees is their pollination service to flowers – wasps do all this too.

Inside the wasp factory

Wasps have an annual cycle that starts in the spring with a single queen. Having overwintered, she sets out to found a colony, which by the end of the summer will be a bustling metropolis of insect activity with several thousand individuals milling around inside a paper nest. Some species nest underground and some have very well-hidden nests in trees and hedges. Safe observation of their to-ing and fro-ing can only be made at a distance with binoculars.

High rise flats

The nest construction and what goes on inside is mind-blowingly beautiful. Once I had a bizarre opportunity to witness the inside of a living wasp nest that had been built up against a window.

Inside were several layers of cells like the floors in a block of flats, but upside down. I could watch the larvae, stuffed into the cells like fat people in sleeping bags, being tended by worker wasps. Whenever the grubs wanted attention they scratched their mouthparts on the sides of their cells, a noise I could hear even through the glass.

The queen was only sometimes evident, but the workers were everywhere. I watched as they flew backwards and forwards with mouths full of wood pulp, which they spat out and crafted with their jaws, every mouthful set and dried like papier maché.

Left: When hibernating, queens take on a very distinctive posture, often gripping with the jaws. The legs and wings are folded up and sit in a groove along the side of the body. Most of these insects do not actually survive the winter; many die of exposure or starvation or are eaten by birds, rodents and spiders.

Cool air system

One of the most remarkable things to watch was the way these insects controlled the internal conditions of the nest.

As the day progressed, the sun moved around the sky and warmed things up inside. Some workers sat in the entrance fanning their wings, to get cool air circulating; others fanned around inside pushing the warm air around further. As things got hotter, some made trips to a nearby birdbath to gather water. They spat this out on the walls and cells so that evaporation would keep the nest cool.

The embryo nest

At first, the nest is usually tended only by the queen. She starts off doing everything herself, building the first cluster of brood cells and enclosing the entire thing in a fragile envelope of carton or wasp paper (wood pulp mixed with her saliva). She feeds and incubates the grubs once they have hatched by curling her body around the base of the cells. If you find one of these strange structures, the first thing to do is find out whether or not the insect is in. Many nests are abandoned at this stage.

The queen may well be out foraging during the day. At night she will return home, so gently approach the nest with a torch and try looking through the entrance hole to see if she is in. If she has abandoned the nest, and you are sure she has not returned, you can examine it close up. Look for different coloured bands of carton on the outside, each representing a different source of wood fibre.

The remarkable thing about these nests is that the carton can be recycled, rewetted and remoulded into new shapes as the colony swells. You can try cutting the envelope in half with scissors to see the brood cells and the structure inside.

Above: A wasp collecting wood pulp for the nest. Look for workers collecting thin shavings from garden furniture, gates and fence posts.

Right: This embryo wasp nest in a loft space is the work of one queen. Once her work force starts to emerge, the nest will grow beyond belief.

The workers' rebellion

The new queens-to-be and the males are released from the colony in late summer. This is the event the whole nest has been preparing for and it is worth looking out for a behaviour known as 'hill topping'. The wasps all head for a high point such as a church spire, tall hedge, tree or high wall.

Back at the nest things start to change. The wasps are no longer the well-behaved workers that toiled so selflessly for their queen and colony just a month or so ago. They are now drowsy and disorganised and can be a plain nuisance to humans.

The elderly queen slows down. Her influence over the workers is diminished. She lays fewer eggs, so there are fewer workers to tend the larvae still in the cells. These larvae are weaker because the male wasps have been feeding on their saliva prior to leaving for their first (and final) fling.

The queen had been maintaining her reign with a chemical that subdues the wasps who tend to her. Without it the workers riot and fight. The chemical usually gets spread around the colony as worker feeds worker and more importantly as worker feeds grub. With fewer grubs and workers the orderly insect palace becomes a right royal mess.

Right: A Common Wasp, one of the most under-appreciated insects in the world.

Below: Workers from a wasp nest binge on rotting fruit, having been released from their nest duties by the fading queen.

Going it alone: solitary wasps

Solitary wasps are very hard to identify, because they come from all sorts of different backgrounds and have a range of appearances. The one thing they have in common is that they are all hyperactive and love sunny situations. The females work alone to construct a nest, usually a burrow or hollowed out plant stem. In it they place a paralysed insect or spider as fresh food for their grubs, which hatch from eggs laid inside the cell.

As an introduction to these exciting insects, find a warm sunny and sandy location – dunes, heaths and sunny path edges are good – and look for holes and winged activity. Identification can be a bit of a minefield but that shouldn't stop your enjoyment of the show.

The frenetic *Ammophila sabulosa* is one of the largest solitary wasps. It is often called the sand wasp, because of its love of sandy soils. These large black spindly wasps with a red/brown waist band can be seen digging, flying off with loads of excavated material, stoppering their burrows to deter squatters before returning with a paralysed caterpillar to stuff in the hole along with an egg.

If you sit still long enough you can watch the nesting process, including the circular orientation flight as she leaves her nest site. This is when she takes in landmarks like stones and plants so she can find her home when she comes back. Move a stone out of place and she will get confused; replace the stone and she will pick up from where she left off.

Other species often share the same colony location. Look out for black and yellow striped weevil hunters of the Cerceris genus, stocking their mini volcanoes withbeetle cargo. And try and find one ofthe cuckoo wasps such as the Ruby-tail Wasp – *Chrysis* species – an animal whose low down and dirty ways (laying its eggs in the nests of others) is masked by its beautiful metallic red and blue garb.

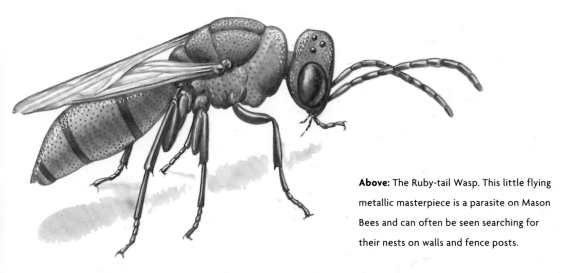

Above: The Ruby-tail Wasp. This little flying metallic masterpiece is a parasite on Mason Bees and can often be seen searching for their nests on walls and fence posts.

Bees

Do you think that a wasp is just a wasp and a bee is a bee? Well, have I got a surprise for you! If you think all bees live in hives and make lots of honey for us, think again.

HYMENOPTERA

Scientists call the ants, bees, wasps and sawflies Hymenoptera. The name refers to their wings. Hymen means membrane and ptera means wing. One look at a bee or wasp and you can see why – most have two pairs of thin, transparent wings.

A sting in the tail

A sting is actually a modified egg laying tube so stings are found only in females. The sting itself is usually a retractable gadget that sits in a sheath at the end of the abdomen. Only the bees, social wasps and some antsuse their sting in defence. A few of the solitary wasps will use it to hunt and to paralyse their prey. All their other relatives use their egg laying tube for its original purpose of piercing and probing various animals and objects to lay their eggs.

Stings are never used in malice! Bees and wasps have to have a very good reason to get their stings out. Using them is expensive to the insect. Wasps have to make venom inside a special venom sac: to use this unnecessarily is a waste of energy. For Honey Bees the cost is even greater because the sting is barbed like a fishing hook: if it goes into anything but another insect it gets stuck in its victim and with it comes a bit of the bee. The bee that dealt the blow will usually crawl off and die of its injuries somewhere.

We couldn't do it without them ...

If all the bees, ants, wasps and sawflies were to vanish from the face of the Earth human society as we know it would start to crumble and pack up. We humans simply cannot live without them! Why? We all know that Honey Bees make honey, which we find very pleasant. In fact, it was the original sugar and has been used by man for at least 9,000 years. A few thousand years later we had started keeping them and domesticating them in hives. But this is not the

Above: The Honey Bee, which will apparently visit 64 million flowers to make just 1 kg of honey!

Left: A bee box in a sunny position can attract many species of solitary bee to your garden.

reason for their importance; bees and wasps have a major role in pollinating flowers, without which we would have no fruit. In fact, most of the plants we rely on for crops are dependent on pollination.

So much so that many species of bumble bee are used in greenhouses and introduced to other countries to perform just this task on our behalf.

Life styles of the socialites

A typical bee that is well known in the garden, especially in spring, is the bumble bee. Actually, there are 19 species of bumble bee living in Britain and 9 species of look-a-likes, which are parasites. Bumble bees are instantly likeable; they have a cuddly appearance and they often turn up in our gardens well before any other flying insects are around, which makes us notice them even more.

The other common social bee is the Honey Bee, *Apis mellifera*. This is a very special insect in many ways. It is thought that it came from somewhere in South-east Asia or Africa and there are drawings on cave walls in China, 9,000 years old, that show humans collecting honey from bee nests.

Honey Bees may have evolved in the tropics but modified their lives just a little in order to deal with colder climates and so spread north into Europe naturally. Or they may have been introduced by humans as they spread out around the globe.

The secret to their success is that whether in the Serengeti or Haywards Heath, bees store honey in their combs to see them through the rough times. This can be for either the dry

Bee bodies

Rather than run away screaming, watch a bee at a flower – it isn't going to sting you. Look for details and features that give you a clue as to how it lives.

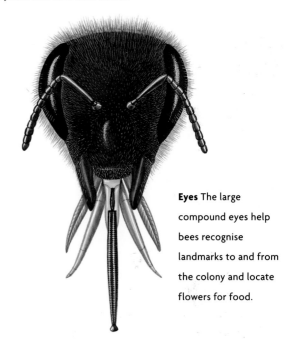

Eyes The large compound eyes help bees recognise landmarks to and from the colony and locate flowers for food.

Mouthparts Watch a bee feeding. See how it spoons up the flower's nectar, rather than simply sucking like a butterfly.

IF YOU GET STUNG

If you do get stung by a Honey Bee do not pinch the sting out with your fingers: by doing this you squash the venom sac and in goes more venom and up goes the pain! Instead scrape the sting away with your finger nail.

season when there are no flowers, or the winter when there are also no flowers. The only difference is the temperature, but bees are naturally very good at maintaining a constant temperature within the nest, by generating more body heat fuelled by this honey store and clustering together as a tight bunch, or cooling down by spreading out, fanning their wings or spreading water around.

Honey bee or wasp?

Thanks to the popular idea of bees as black and yellow stripy things, we tend to grow up with the wrong image in our minds. I have seen many people spot a Honey Bee and immediately call it a wasp.

The simplest way to tell them apart is that Honey Bees are a sort of orangey brown (a similar colour to the runny honey they produce) and quite hairy. Social wasps are very bold yellow with black bands on their bodies.

Temperature control

Bumble bees have very fat and furry bodies compared with other members of the bee family. They are able to get their nesting and breeding underway a lot quicker in the season because of this and a few other neat adaptations.

The hairy coat keeps the insect warm. By sitting still and disengaging its wings from its flight muscles (like taking the wings out of gear), the bee can vibrate these big muscles very quickly. This is like running on the spot – not going anywhere, but getting hot. The bumble bee is warming its body up, either to get ready for active flight or to incubate her brood. Like a bird she sits hunched over her eggs. Even when there is a late spring there could be a queen bumble bee deep below the surface of the soil, snug in her insulated nest, generating a temperature of 30–32 °C (the same as a very hot summer's day). Being able to do this means the

Wasp

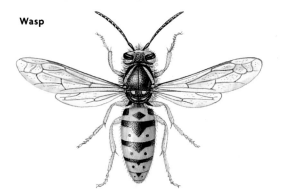

Bee

Wasp or bee? There is no need to fear either of them. Social wasps have very yellow and black stripes, while honey bees are fluffy with ill-defined stripes of brown and orange.

LOOK OUT FOR

Make your garden wasp- and bee-friendly by planting lots of nectar- and pollen-rich flowers, which are also highly attractive tobutterflies and moths. You can also make nest sites for bees and wasps, in the same way as you might provide boxes for birds.

Right: Catkins of Pussy Willow are one of the best places for Honey Bees to stop at in early spring when there are few supplies of pollen and nectar.

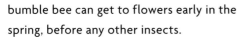

bumble bee can get to flowers early in the spring, before any other insects.

Honey Bees are absolute masters of controlling the temperature of their nest. During the winter months workers cluster together around their queen in the very centre of the nest. Here, fuelled by the store of honey, they can generate their own heat. If it gets colder, they cluster tighter together. The temperature in the centre rarely falls below 17 °C and more often it is close to 30 °C. When spring arrives, the bees can go out and forage if conditions are good, otherwise they just stay in the nest and eat honey. They keep going like this year in year out, for as long as the queen survives in the nest.

Bumble bee lifecycle

Bumble bees have an annual cycle, unlike Honey Bee colonies. They start and finish the whole colony life in one year. One of our commonest and most widespread bumble bees, *Bombus terrestris* is also sometimes called the Buff-tailed Bumble Bee. It starts its lifecycle early on a warm spring day. As soon as the soil has warmed up, the big furry bundle wakes up. The queen has been deep in a crevice in the soil all winter. For five or six months she has been living off the fat reserves in her body. She has also stored sperm after mating the previous summer.

When she wakes up, she will feed on nectar and pollen from early flowers and then begin investigating cracks, crevices and dark places.

Stocking up

The queen selects a suitable nest site, usually an old mouse nest, although it can be pretty much any kind of cavity as long as it is dry and well insulated. She will stock it with a pad of pollen which she collects in special hairs called baskets on her hind legs. She then models waxy secretions from her body to form a rough cell around the pollen pad into which she lays her eggs. At the same time she builds herself a little wax cup and fills it with nectar or honey. When the larvae hatch, she rears them on a mixture of pollen and nectar. Eventually, the larvae pupate in a cocoon of silk.

When the weather is good in spring, look out for these new queens. They are easy to recognise by size. If you see one collecting pollen (in two large yellow lumps on her hind legs), you know she is provisioning the nest.You can follow these queens to find out where the nest is.

From one to many

Look out for the first generation of little bumble bees flying around in spring, reared single-handedly by the queen. Once enough workers have hatched out to take over, the queen stops foraging herself and concentrates on egg laying and cell building. The colony can swell to several hundred bees in the nest.

MAKING A BEE BOX

This is a neat and simple way of creating a perfect place for solitary bees and wasps. Many species will hole up in the wild in hollow stems and the excavations of other insects, especially those left by wood-boring beetles. You can have a go at recreating these conditions by making a bee box.

1 Collect old hollow twigs and plant stems of different diameters, such as the dead stems of hogweed, bamboo garden canes, buddleia stems, elder twigs, grasses and reed, and even paper drinking straws.

2 Now fill a container such as a plastic flower pot, baked bean tin, or the bottom half of a plastic drinks bottle with these stems. Trim them so they are more or less flush with the rim of the container. You can also make a container out of wood, like the one in the photo.

3 Attach it to the stem of a shrub, a tree trunk or a garden cane with string, wire or twine. The advantage of using a garden cane is that you can move it around.

The entrances should all be sloping down slightly. This stops them getting waterlogged when it rains. Place the nests in a sunny, open location, preferably south facing – these insects like it hot!

Ants

Sadly, we normally react to a nest of these insects in one of two ways: we reach for the insecticide or boil a kettle to douse the unfortunate colony! Why? Are we subconsciously threatened by the mighty empires they run? Are we a little jealous of their immaculate house keeping?

There are around 50 different kinds of ant in the British Isles and if you tried to identify any of them you would probably end up going mad.

I strongly recommend making a formicarium. It may seem like a lot of hassle but it really does open up a window on the lives of these bustling little animals. I hadn't kept ants since I was 11 years old, but for this book I thought I would get a colony going for a bit of inspiration ... wow! I'm an ant addict again!

Just what is an ant?

Ants are small wingless wasps that have been busying themselves on planet Earth for the best part of 80 million years. They are one of the most successful groups of animals on the planet today. If the Earth belongs to anything, then it is Planet of the Ants!

Ants are fine weather insects, not liking the colder climates very much, and compared with the rest of the world Britain has a very sparse selection of the 9,000 or so species that exist.

They live in the same sorts of colonies as the highly social bees and wasps with a queen or several queens depending on the species. The queen lays all the eggs and she also controls the colony's actions with various smelly chemicals called pheromones. In this way each nest behaves rather like a single animal and each ant as a cell.

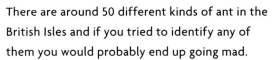

Above: These Wood Ants are giving this ladybird a hard time; perhaps it was eating their herd of aphids. In the same way a shepherd protects his flock, ants care for their aphids.

Left: In spring, look under stones to find ant nests. Ants use the stones like storage heaters; the stones warm up in the sun and release this energy slowly to the ants below.

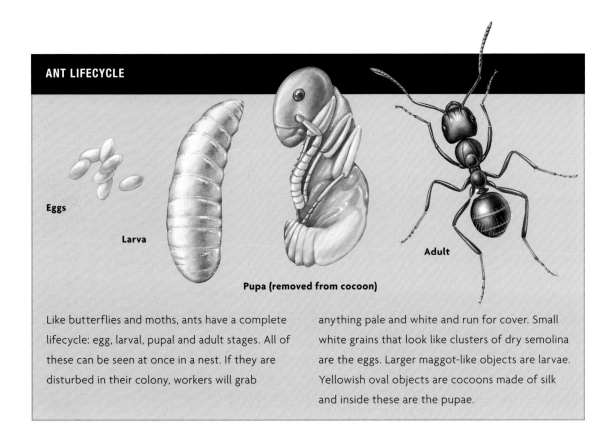

ANT LIFECYCLE

Eggs

Larva

Pupa (removed from cocoon)

Adult

Like butterflies and moths, ants have a complete lifecycle: egg, larval, pupal and adult stages. All of these can be seen at once in a nest. If they are disturbed in their colony, workers will grab anything pale and white and run for cover. Small white grains that look like clusters of dry semolina are the eggs. Larger maggot-like objects are larvae. Yellowish oval objects are cocoons made of silk and inside these are the pupae.

From colony to colony

When ants go out on foraging trails they become the eyes, nose and arms of the colony, searching and sniffing out food and other resources to pick up and bring back to the nest.

One of the commonest, most widespread and familiar of the British ants is the common Black Pavement Ant, *Lasius niger*. This species is a good model for a run down of the life of a colony.

Sexodus

In mid to late summer thousands of winged ants erupt from the cracks in patios, tarmac and paving slabs. Each colony has been pampering the males and potential future queens.

Ant colonies you have hardly even registered spill out. You may not see them, but on some environmental cue – it could be day length, temperature or humidity – every ant nest in the neighbourhood will be engaged in the launch, ensuring that the maximum genetic mixing occurs. They fly out, spiralling up into the air to mate in the air, in trees and on roof tops.

The queens return to the ground, and get rid of their wings by twisting, rubbing and pulling at them with their jaws. They will never fly again.

The males die; their task is complete. The new queen now has enough sperm inside her body to last her a lifetime.

Single mum

A new queen may be adopted by another nest or even the one she just left but more often than not she will start her own empire.

She finds a small hole, crack or cavity in the soil or under a rock and seals herself in. This is known as her claustral cell. She then lays her first eggs and rears the grubs or larvae that hatch from them to be the first workers of the colony. Until the young are ready the queen will not emerge into the outside world, not even to feed. She relies on fat reserves in her body and the big wing muscles that powered her first and only flight.

Colony rules

Once the queen has her first set of workers, she can sit back and be pampered and fed by her daughters and get down to the serious business of laying eggs for the rest of her life. The queen lays more and more eggs, and with more and more helpers being produced the colony just gets bigger and bigger, expanding throughout the next year or so.

Lift a stone or peer into a formicarium and you will see just how organised things are. There are chambers containing eggs, larvae of different ages and sizes and the pupae.

You will notice that the ants are all busy but if you can follow a few individuals with your eyes, you will see that they tend to stick to certain jobs. They carry out a number of different jobs during their lifetime, a kind of job promotion scheme.

An individual might start as a nurse ant, looking after the broods of eggs and larvae, constantly cleaning them, feeding them and moving them around to the part of the nest which has the best humidity and temperature for them to grow up in.

Then he or she might get to work on the nest, building extensions and making repairs (ants are rather haphazard builders compared to the bees and wasps).

Finally it would be time to forage, returning to

NUPTIAL FLIGHTS

Lots of insect-eating animals cash in on the superabundance of protein when ants undergo their nuptial flights. Look for newly fledged Swallows and House Martins, Starlings, Sparrows, even Black-headed Gulls, all feasting on flying ants.

Right: Every year, usually on a hot, August day, Black Pavement Ants erupt from their nests. Winged males and future queens engage in a mass mating.

the colony with a crop full of food to feed to all those ants that do not get out.

Once the nest reaches a certain size, the queen allows for the production of more virgin queens and males. These are produced in the nest during the summer and here they wait for those muggy, warm conditions for the big mating flights that start the whole thing off.

This cycle continues as long as the queen lives, which can be as long as ten years.

What do ants eat?

Stake out a wood ants' nest and watch one of the trails leading to it. The trail of ants moving

toward the nest will be carrying all manner of goodies: worms, flies, caterpillars and aphids, some in pieces, some dead and dying. This is what the ants are feeding their larvae on, protein rich and packed with goodness, perfect for the developing grubs.

Most species feed on a mixture of both meat and plants. Some rely on certain sources more than others, for example some ants feed on protein-rich seeds. Ants also eat the occasional flower for a sugary supplements to their diet. But much of the vegetable part of their diet comes from the sweet sticky honeydew produced by aphids which the ants 'milk'. There

Above: The hustle and bustle of life in the ant colony can be revealed by lifting a stone. Here you can see workers scurrying off with larvae.

Above: Fierce carnivores, ants will hunt down, kill and remove to the nest any animal that cannot defend itself, even prey much bigger than them.

are aphids which ants do not seem to milk and nobody really knows why this is, as they also produce honeydew. And the relationship between ants and aphids can break down from time to time, especially if conditions become too crowded in the herd or it's getting late in the year. When the time comes the ants cull their stock and take the aphids back to the ever-hungry grubs.

Recruitment and communication

One reason for ants' success is their efficient communication system. They use scent in a very clever way. Have you ever looked down and seen an ant randomly walking about on the patio? For every one ant on the patio, there'll be another on the wall, a few in the borders, a couple around the dustbin, several on the lawn and so on.

Some species get into our kitchens through the tiniest crack. All it takes is for one to get lucky and find a pile of desirable food and the secret of ants' success will slowly unfurl. Next time you see an ant at home, place a dollop of jam in front of it and sit back and watch.

To start with the ant will probably fill both of its stomachs. The first one is a crop and this contains the food that is often shared around the colony, a kind of community stomach. But the ant is still carrying only a little food and the jam represents a mountain of useful resources which the single ant has not even dented.

It will then run back to the nest, dragging its abdomen along the ground. It is leaving a chemical trail that is dribbled from glands in its abdomen. This will start to evaporate quickly. Back at the nest, other workers are greeted with lots of excited antennae-waving and they will set off along the trail, smelling the vapour as it wafts up from the ground.

Soon this is repeated by lots of ants, each reinforcing the scent trail. If you watch your blob of jam for long enough or even revisit every few minutes, you will notice just how effective their numbers are. As the food is used up and taken back to the nest, the number of workers returning decreases, the concentration of the scent trail peters out and no more ants return.

ANT FARMERS

If you watch ants visiting a greenfly colony on garden plants you will see them stroking the little bugs in a way reminiscent of a person milking a cow. Instead of milk the aphid delivers a little drop of honeydew, plant sap that has passed through the bug's body. The ants go crazy for this stuff. Some ants move the aphids around to new pastures, and certain ants even build little 'sheds' of soil to protect the aphids. This is taken to extremes by ants that farm aphids that are feed on the roots of plants. These ants build their own nests around their aphid herds.

Making a formicarium or ant city is the best way to get to know ants, as the secret of their society is usually underground, or at least under a stone! It's best to start this in early spring. It is critical you follow the instructions and get the measurements right otherwise your ants might escape.

YOU WILL NEED:
plaster of Paris • jug • jam jar with lid • spoon • plywood for base • 6 strips of wood approximately 2 cm x 2cm in cross section: 1 x 30 cm, 2 x 20 cm, 3 x 12 cm • modelling clay • clear plastic tube, at least 30 cm long • sheet of hard clear plastic • black cloth or paper • ants with soil • duct tape • wood glue

1: To make the formicarium you need to glue the frame of wood strips to the base as in the photo. With the two of the 12 cm long pieces, make a wall with a gap in the centre (this will be the entrance). Just beyond this gently place the remaining 12 cm strip of wood. Do not glue it down because it is part of the mould and will be removed.

2: Next, using the modelling clay, make a system of tunnels and chambers, as deep as the frame. The space taken up by the modelling clay will eventually be the living space and corridors for your ants. Completely fill the entrance with the clay and ensure the modelling clay tunnels butt up to the loose wooden strip. Place the lid of hard clear plastic on top at this stage; it helps you to see whether the modelling clay is high enough. If you have got the height right the clay should squich gently against the plastic.

Leave a plug of modelling clay in the entrance, and push one end of the clear plastic tube through it. The end of the tube in the nest should also butt up tightly to the loose piece of 12 cm wood as in the photo.

3: Mix up the plaster of Paris in the jug into a runny solution and pour it into the gaps until it comes to the top of the frame and fills all the spaces. Leave it to set for a day or so. Then pull out the modelling clay and loose piece of wood. You will be left with a series of tunnels in the plaster of Paris.

4: Fill the cavities left by the modelling clay with well drained soil. Make a hole in the lid of the jam jar, large enough for the plastic tube to fit through, and attach the jar to the end of the entrance tube. The tube must be flush with the inside of the lid, and you should seal any

gaps with more modelling clay. Place the hard, clear plastic lid on the nest and secure and make completely ant-proof with duct tabe.

5: Collect some ants. The Black Pavement Ants are best, although Red Ants can be used as well. Look for the their nests under stones, tin and wood. They are best collected in early spring as they cluster together and can easily be scooped up. For your colony to thrive you need to be guaranteed a queen and a good selection of workers, eggs and pupae.

Introduce them into the jam jar and add a little cotton wool soaked in weak sugar solution. Then cover your formicarium with the black paper or cloth, leaving the jam jar exposed to the light. This gives your ants a nice dark atmosphere to set up their new ant city and will encourage them to leave the jar for the formicarium.

To start with the ants will seem in complete turmoil but soon they will organise themselves and start to move into the formicarium. Leave them in the dark for a few days and let them become established. You can now add different food types in the jam jar. They require sugar, water and protein, which you can provide with honey, damp cotton wool, insects (collect these alive and dead from window sills) and seeds.

Assuming you got a healthy queen when you collected the ants, you will soon notice the numbers start to increase as the workers rear more recruits and the queen lays more eggs. Make sure you keep the whole formicarium in a warm location but out of the way of direct sunlight and provide the ants with plenty of food and water – you can simply substitute the jam jar with a new one when you need to clean it out.

Look after these basic needs and the ants will give up many of their secrets. Every day you will notice something different. Enjoy observing them – you will be amazed.

CAN YOU REVEAL THE SECRET?

Try to persuade an ant returning to its nest to walk over a piece of glass or perspex. As soon as it has, sprinkle some talcum powder on the glass and blow the excess off. You might just see a faint line of talc where it has stuck to the liquid the ant has squeezed out of its abdomen. Have patience: some ants give better results than others.

1 Make the frame.
2 Make a system of tunnels and chambers.
3 Mix up the plaster of Paris and pour into the mould.
4 Fill the cavities with soil.
5 Add your ants.

Bugs

Just flick open the bug section in a good field guide and what you see before you is an amazing array of beasts – furry ones, funky coloured ones, sap-suckers, blood-suckers, crawlers, hoppers, fliers, swimmers and even a few that walk on water. There are over 1,600 different species just in Britain and Ireland!

A bug's body

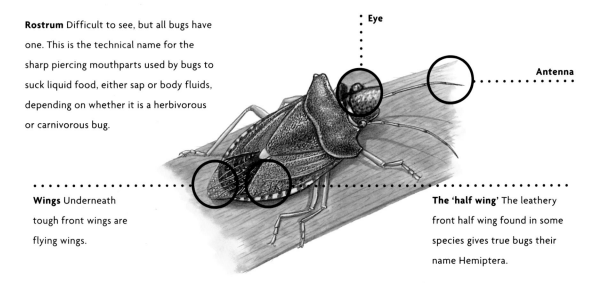

Rostrum Difficult to see, but all bugs have one. This is the technical name for the sharp piercing mouthparts used by bugs to suck liquid food, either sap or body fluids, depending on whether it is a herbivorous or carnivorous bug.

Eye

Antenna

Wings Underneath tough front wings are flying wings.

The 'half wing' The leathery front half wing found in some species gives true bugs their name Hemiptera.

What is a bug?
The technical name for the family is Hemiptera, which means half wing. This refers to the distinctive half leathery and half delicate wings of some species.

The one thing that all bugs have and which makes them different from beetles and all other insects is their sharp needle-like mouthparts.

All the true bugs, from herds of greenfly to pretty shield bugs and pond skaters, use those sharp mouthparts but for very different jobs. Some use them like a straw, plugging into succulent plants for their juice; others use them like a dagger, for stabbing prey or in defence.

It's a bug's life!
Bugs grow up without a pupal stage. On hatching from the eggs they look just like their parents, just smaller and without wings. Then they start eating.

They grow, as do all insects, by shedding their skin, usually four times, until they reach adulthood.

In the wacky world of the aphid things

ESCAPE TACTICS

Despite looking completely defenceless, aphids have a number of escape tactics. Tickle a colony with a paint brush. You will see a lot of the aphids simply let go of the twig or bud on which they were feeding. Some will also present their rear end to an attacker and ooze a waxy substance out of the two prongs – called siphunculi – either side of their abdomen. This is sticky and unpleasant especially as it solidifies on the mouthparts or the eyes of a ladybird. The odour of the wax acts as an alarm call to other aphids which simply unplug their mouthparts from the plant and walk or drop off.

are taken a step further. Let's take the common Blackfly as an example. It sits out winter as an egg, secreted away in a crack of the bark of a plant such as the Spindle Tree or Guelder Rose.

As spring warms up, the eggs hatch and the young nymphs plug into the unfurling sap-rich shoots. Four moults later they are plump little females, all of which begin to produce clones of themselves. This continues until clusters of Blackfly adorn the shoots and buds, reproducing faster as temperatures rise.

Then there comes a trigger – probably overcrowding. If they were to stay, the plant would be drained and they would eventually run out of sap. So rather than go hungry, a flying generation is born with different tastes. These winged individuals fly off to spend summer on other plants such as docks, thistles and beans. They then settle down and start breeding and feeding all over again, until the days shorten.

Then a generation of winged females is born. They fly back to the winter host plants and give birth to a generation of egg-laying females. Back on the summer host plants, the shrinking numbers give rise to winged males. These fly off and seek out the egg laying females in the few hours that they have to live as adults. After mating the females lay tiny eggs in the bark ready once more to see out the winter. Talk about complicated!

BUG LIFECYCLE

Once the eggs (1) hatch, a miniature wingless version of the adult emerges (2). Through a number of skin sheds it eventually develops into a fully grown big bug (3).

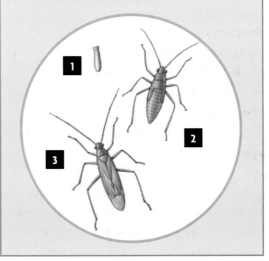

Experiment: Bubble bum

Every spring little gobs of foaming spit appear, seemingly splattered around in our gardens and hedgerows. This is cuckoo spit, but it's not how it sounds... this is actually the clever concoction of the young stages of the common green Froghopper. The bubbles are a kind of camouflage to protect the young nymph. .

YOU WILL NEED:

fine camel hair artist's paintbrush •pot of water

Wet the paintbrush and gently dab at the bubbles, wiping them off until the little green insect is revealed. It will probably be aware its cover has been blown, and will hide on the other side of the plant stem.

If you leave it now, it will plug its sharp straw-like mouthparts back into the stem and before your very eyes you can watch it blow a new cover from its bottom. The

Right: Cuckoo spit is nothing to do with that early spring bird, but a defence whipped up by a young Froghopper.

bubble liquid is actually the plant sap that has passed right through the insect.

Later on in the spring you may stumble upon froth of a slightly different consistency. This is the cocoon from which the nymph makes its final moult into a winged adult.

No dads required

Anyone who has tried to grow plants will have encountered the incredible potential of aphid populations to expand.

Try taking a stem with only a few Greenfly or Blackfly on and place it in a vase indoors. Select a stem without any winged animals; you do not want to introduce these guys to your house plants.

Keep an eye on the big ones; these will be the adult females and assuming you are in the middle of summer, they will be in full swing.

As well as excreting globules of plant sap

called honeydew, they pop out the occasional baby aphid – no male required, no time wasted with courtship. They just churn out lots of little clones.

This is a rare occurrence anywhere in nature, but aphids are rather good at it. Some species can turn out up to ten young a day like this. Each of these mothers is not just giving birth to daughters, but inside these daughters are fully developed embryos of their granddaughters too. If all of these survive, the single female can give rise to over 6 million aphids in two months!

A bug for every occasion

There are so many different bugs, in so many different shapes and sizes that there is no way I can show them all here. Here are a few that you might bump into in your back garden or park.

Shield bugs

There are over 40 different kinds of shield bug, some of the most spectacular and largest of the land-loving bugs. Most are flat and quite a few are shield-shaped.

The other name for them is stink bugs – pick one up and smell your fingers afterwards. You have just been bugged! They ooze a very smelly distasteful chemical from glands in their thorax. Some advertise themselves with bright colours as a way of warning any animal stupid or tasteless enough to try eating one.

Water bugs

The water bugs are a motley bunch and not all of them can actually swim very well. They love densely weeded pools and one of the highlights of any pond dipping session is to peer into a weed-filled net and notice one of the plant stems get up and start walking off. This is the Water Stick-insect – a bad name and a good name all in one. Yes, it looks like a stick insect, but it isn't even closely related to one! It is yet another peculiar bug. Big eyes, stout beak and big front legs a bit like a Praying Mantis, just waiting to stretch out and give a death hug to any passing small fish, tadpole or insect. The big pointy thing sticking out of its back end is not a sting as is often claimed, but no more than a snorkel, which draws air from the surface, and stores it under the wings, where it is absorbed by the spiracles on the abdomen.

A similar but stouter looking beast is the equally badly named Water Scorpion. This animal has all the above features but in a more compact style.

The other more active species include the frisbee flattened Saucer Bug, the Backswimmer and the Lesser Water Boatman. Peer into any water body bigger than a puddle

Top: Hawthorn Shield Bug. It's easy to see how the shield bugs got their name.

Above: The Water Measurer's slow lazy pace is unusual for a predator, but this bug's dinner is usually trapped firmly in the water.

during the summer and you will more often than not see one of these rowing away to the bottom. These are bubble breathing bugs. They trap a layer of air from the surface around their bodies, usually between their wing cases and their abdomens and on hairs on the underside of their bodies too.

This portable air supply, which takes on a silvery appearance if you look at one in a jam jar or tank full of water, allows the insects great freedom under water. They are some of the fastest aquanauts in the pond. The Backswimmer and the Saucer Bug are both voracious predators, using their mouthparts like a dagger, stabbing and eating anything that is small enough for them to manage.

The Lesser Water Boatman is the only vegetarian among those mentioned. It uses its rostrum like a vacuum cleaner, sucking up algae and plant debris that it has sifted through with its front legs.

The aeronauts

If you make a new pond in your garden or see a bundle of water bugs happily rowing around in a cattle trough, have you ever wondered how they get there? There is a very simple explanation.

To see it in action, you need to catch an adult insect using a pond net. A Water Boatman is ideal – they are the ones with the silvery white back, easy to see when they are flipping around in your net.

Gently lift the insect from the net (try not to squeeze it as Water Boatmen can deliver a painful bite). Hold it in the flat of your hand; it may leap about a bit at first. Eventually it will settle down with its back facing up. Now is the time to watch carefully. In my experience the next bit usually happens in a couple of minutes, but it may take longer. It depends on things like how sunny it is and the air temperature.

Watch for it to pop the top of its back open like the bonnet of a car and start pumping its abdomen back and forth. This is the sign. The animal will unfold a set of wings and buzz off to find a new pond.

COURTSHIP SERENADES

The Lesser Water Boatman is one of the few pond insects to use sound in its courtship. If you keep a few of these insects in a tank for observation, keep your ears open for a squeaky, rasping song, particularly noticeable during the quiet times of the day of evening and morning. This sound is produced by the male, which rubs his bristly front legs on a ridge on the side of his head.

Beetles

More than a quarter of all animals on the planet are beetles: over 350,000 different beetles have been named so far. So what makes them so successful? Well, the beetle is a design classic, which has been around for over 230 million years. The secret is in its simplicity.

Perfect cover

Go out to the garden and give a bush a good tap. I bet the majority of animals that fall out of it are beetles. In fact, you could turn upside down any habitat on earth except for those in the polar regions and you would find lots of beetles.

One of the secrets of beetles' huge success is the wing cases that most have on their back. These highly modified forewings, called elytra, are perfect covers for the delicately folded rear flying wings.

Some beetles do not fly and their elytra are fused together – very handy as a nice shiny coat of armour to give excellent protection from predators and the environment.

To see for yourself how the elytra and the wings are used in flight, next time you are in the garden just persuade a ladybird or a Chafer beetle to climb onto your hand. The elytra spring open and after a pause the transparent hindwings are carefully unfolded on delicate micro-hinges and briefly stretched before the beetle goes humming off into the heavens. Beetles may not be the most graceful fliers, but they still have all the advantages that a life with wings offers.

Demons in the dust

Blisteringly hot June days are the best times to go stalking tigers... Tiger Beetles that is. Usually, you'll first notice their shadow skittering quickly over sun-baked ground. Their good vision, thermally charged bodies and long legs mean they remain three steps ahead of you all the time. If you get too close they leap into the air, spread their metallic green wing cases and buzz off to safety.

Watching them hunting on sandy heath, dune and moor is easiest with a pair of close-focusing binoculars but if you team up with a friend you can drive the beetles along a path toward you and if you wait lying on your belly, you may just get to stare one of these grassroot terrorists between the eyes.

To really appreciate a Tiger Beetle's metallic finery and evil-looking mandibles at close quarters, you need to catch one and view it

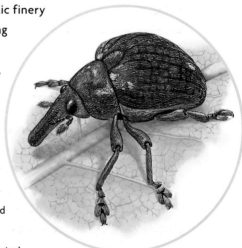

Right: Seen up close under a hand lens Weevils are oozing with character!

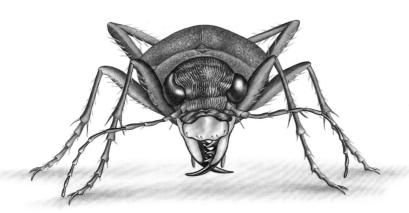

Above: The Tiger Beetle isn't called a tiger for nothing! This is the last thing you would want to see if you were a soft-bodied insect.

Above: The Tiger Beetle larva is a ruthless killer like its mum and dad. But it is an ambush specialist, not an active hunter like its parents.

in a clear plastic pot. The best way of catching them is to use a butterfly net.

If your morbid curiosity still isn't satisfied, look for small, neat holes about 5 mm in diameter in sun-baked mud or sand at the edge of paths. Use a mirror or torch and you may reveal one of the strangest creatures you are ever likely to see through a lens: a Tiger Beetle larva. This ugly bug has a serious appetite and a pair of jaws to match its parents', but unlike them it waits in ambush for its prey. With its all-round vision, it truly has eyes not only in the back of its head but at the sides and in front too.

You can gently exhume one of these bizarre little grubs by carefully excavating the burrow with a blunt knife or spoon. When you have finished looking pop it back into a similar sized hole made with a cocktail stick or skewer.

Little ladies

You may not know much about beetles, but I bet you know what a ladybird looks like and would recognise one instantly.

You might also know they are good for the garden because they chomp aphids. But did you know that ladybirds start their career in pest control even before they take on their adult form and you can find their babies fairly easily?

In June or July look for a plant covered with aphids; stinging nettles and thistles often have a good population. Look closer still and you will almost certainly find a strange beast that looks a bit like a caterpillar – mostly grey but with yellow/orange spots and bristly warts, and a long segmented body.

Give it a tickle and it will sprint off, not in caterpillar style but propelled by six long spindly legs. This is a ladybird larva.

Ladybirds are very easy to rear in a clear plastic box lined with a piece of kitchen towel to absorb excess moisture.

YOU WILL NEED:

clear plastic box • kitchen towel • ladybird larva • aphids • leaves

Put no more than ten ladybirds in each box and try to match them in size. Ladybirds are topnotch predators and if the supply of aphids runs low or they are overcrowded, they will turn cannibal.

Make sure that you clean the pots out every day and provide fresh aphids on leaves and you should get on just fine. You will be able to watch as these voracious larvae gobble up hundreds of aphids.

Look out for skin changes as they grow. There are usually four of these. Just before the larva moults into the pupa, it stops moving, attaches the tip of its abdomen to a surface and hunches up. Do not panic, this is normal behaviour. It is called a pre-pupa and 24 hours later it will have moulted for the last time.

The pupa is a strange mobile thing, which will stand up and down several times if it is touched. It is usually the same colour as the larva, but is still not looking much like a ladybird. About two weeks later a plain-looking beetle will crawl out. It will rest for a while, allowing its wing cases to harden and within a couple of days, colours will come to its wing cases and it will finally turn into the ladybird we know and love.

Catching beetles

As night falls the big beetles emerge from their lairs and go on the prowl. These are the top predators of the insect underworld ready to run down some unfortunate invertebrate and tear it limb from limb. To a slug the ground beetles are every bit as formidable as a Lion would be to a Gazelle.

You can find some of them by looking for their hidey holes by day, turning over logs, stones and debris. Or try the lazy beetle hunter's technique – a pitfall trap. Any old canister with deep, smooth slippery sides will do, the idea being that you sink this into the ground, the beetle falls in and cannot crawl out.

Top: This ladybird is emerging from its pupa. It would be quite a rare sight in the wild, but you can rear ladybirds at home.

Above: 35 spots! A cluster of five Seven-spot ladybirds.

Experiment: Making a pitfall trap

The most basic version is a tin or jam jar sunk into the ground, with a few dry leaves in the bottom and a stone or tile placed over the top, balanced on a few smaller sticks or stones. You can place these anywhere in the garden and hope that a beetle falls in. This works well enough, but the best trap is a little more sophisticated.

YOU WILL NEED:

three sticks • string • 1 large and 1 smaller plastic drinks bottle • skewer • scissors • leaves and twigs • trowel • very ripe bananas • sugar

1: Take a large drinks bottle and cut the neck off it with the scissors. This is going to be the pit. Using the skewer, make plenty of drainage holes in the bottom of it. This is very important because if the traps fill up with rain water the beetles will drown.

2: Dig a hole deep enough to sink the pit and place the trap in the ground so that the lip is level with the soil around it.

3: Take a smaller drinks bottle. With the skewer make a number of holes in the top two thirds of the bottle.

4: Bananas are one of the best baits, especially nice squishy, smelly ones. Before you ask, I have no idea why carnivorous beetles are attracted to the smell of rotten fruit, especially one they would never have come across in the British countryside! Squeeze the soft banana into the second bottle, add a bit of sugar and leave in a warm place for a few hours.

5: Back at the trap site, make a wigwam out of the three sticks and tie them together at the top. The wigwam must be high enough to allow you to suspend the small bottle with the smelly bait in it over the top of the trap, without it touching the ground. Use string to tie the bait bottle to the sticks and place this over the trap. Put some dead leaves, twigs and bark in the bottom of the trap, but not so much that any insects can use them like a ladder and escape. You want them to fall in then hide themselves away from view of predators and be safe until you find them.

6: Wait until the morning to see what you get. The more traps you make the more beetles you catch and the more you will learn about what prowls around your garden. Check the traps at least daily and always first thing in the morning. Some of these beetles are so mean that if they are left in the traps for too long they well snack on their trap mates.

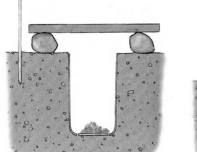
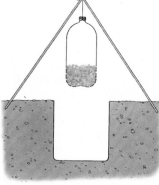

The basic design a pot sunk in the ground is the trap, the tile and stones allow beetles in but stop it filling with rain. Add drainage holes in the bottom. The flag will help you find your trap again the next day.

The beetle's knees a deluxe pitfall trap, worth that little bit of extra work. Don't forget the bananas!

So many species, so little time

There are enough interesting beetles to fill a hundred books like this. Here are just a few of my favourites.

Devil's Coach-horse

The Devil's Coach-horse has always attracted attention. It was once thought that when it raises its tail in alarm it is casting a spell or curse of the devil on you! What's really happening is that you have scared the living daylights out of this normally secretive animal. The tail also emits an unpleasant chemical from organs on the tip. If you pick one of these beetles up with bare hands it will give you a nip with those fierce jaws.

Maybug or Cockchafer

Fizzzzzzzzzzzzzz clunk! That noise heralds the arrival of one of these stunning beetles at your window pane. From May onwards these charismatic beetles will often be attracted to lights and are a common catch in a moth trap during the summer months.

You can check if you're looking at a male or female by counting the number of veins in their fan-like antennae; males have seven, females six. The larvae are white C-shaped grubs that feed on grass roots for three years. These are eaten by birds such as rooks, hence their other name of rook worms.

Stars in the grass

The sight of a Glow Worm often conjures up an image of balmy summer evenings. But how many of us see one and feel sorry for slugs and snails? Find a glowing female and what you are looking at is a murderess.

She's off her food now (she doesn't eat as an adult beetle) but for the last 20 months she will have stabbed, poisoned and consumed many molluscs.

Look for females on unimproved grassland on chalky soils and then search under stones and logs – the sort of habitat in which you might expect to find their prey – for the equally interesting larvae. They look a little like a flattened millipede no more than a centimetre long, black with a pale corner to the rear of each segmented plate.

If you find an adult female glowing away in the grass, look at her with a torch. I know this spoils the spectacle but you will be able to see how un-beetle like she is and also check the surroundings for any males that have been lured in. These look a bit more like the classic beetle, with wings, elytra and a pair of huge eyes, ideal for spotting the females' light show in the grass below.

Left: The Great Diving Beetle is, as its name suggests, found in water. Check your net the next time you're pond-dipping.

True flies and look-alikes

There are several groups of unrelated insects buzzing around that get a little bit confused with flies. It doesn't help that a lot of them have the word fly in their name despite not being true flies at all, such as dragonfly, mayfly and butterfly! Here are a few of the more interesting flies and their look-alikes you might come across on your entomological forays.

A Blowfly body

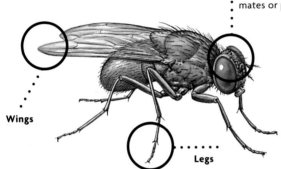

Eyes Many flies work with visual cues. They look for food in flowers and search for their mates or prey on the wing.

Mouthparts These are probably the most variable features of any fly. Flies can do everything from crush pollen, suck nectar and pierce skin to suck blood, stab insects and mop up liquid. Each job requires a different kit of mouthparts.

Wings

Legs

True flies, in particular those Blowflies that we love to hate so much, are actually the most advanced insects.

Winging it

You can see the wing structure quite clearly in the craneflies. Towards the abdomen on the thorax is a tiny pair of hair-like projections with knobs on. Despite being so small they are just as important for flying as the other wings. These reduced and specialised wings are called halteres and they act as balancing organs, helping the insect remain stable in flight and perform those crackpot aeronautics that flies are so famous for.

Below: Cranefly halteres. Look closely at any true fly and you will see these little knob-like organs in place of a proper set of wings. These act as highly specialised balancing organs or gyroscopes.

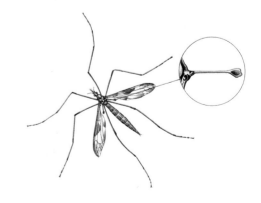

Suck it up

Nearly all flies feed on liquid lunches, whether that is nectar, sweat, urine, the juices from putrefying flesh, or blood, and they have a range of hardware to match these dietary demands.

Next time a Blowfly lands on you take a good look at its mouthparts. They are like a sponge and can mop up anything; if it's a bit dry the flies will spit on their food to help dissolve it and make it more edible.

Different hoverfly species have different mouthparts to reach into different depth flowers, while mosquitoes and horseflies have sharp piercing mouthparts and you know exactly what they do with those – bite!

Munching maggots

All flies have a complete lifecycle. Baby flies look nothing like their parents. We all know these larvae by their popular name of maggots and just like the caterpillar stage of butterflies and moths they are the growing stage, designed to put away as much food as possible.

Mosquitoes – life begins in a pond

Look on the surface of stagnant water in summer for a tiny, grey curved raft about 5 mm across. This button of life is a clump of mosquito eggs and if you collect them and some of the same water, you can watch their lifecycle unfold in a jam jar.

The eggs are laid in little floating boat shapes of several hundred per clump and they float this way up because they have a coating that repels water on all but one end. After only a few days these hatch into tiny, big-headed larvae.

The larvae feed on microscopic debris by using feathery mouthparts to waft food into their mouths. In three weeks or so, after moulting their skins several times, they will eventually change into J-shaped pupae. They can still wriggle around and are highly active, with two snorkels that allow them to breathe while floating under the water's surface.

This stage can last as little as three or four days depending on the water temperature. The adult emerges very quickly as it has to avoid predators.

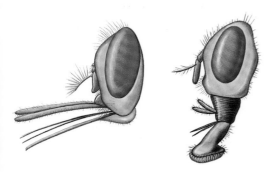

Above: Between them, flies eat anything. Here are the sharp piercing, sucking mouthparts of a biting fly and the sponge-like mouthparts of a Blowfly.

Right: Look in any water butt and you will see a mass of wiggly mosquito larvae

Left: Catching flies in a sweep net.

Flies are fab!

Flies are another insect group that people find it hard to love, but they do a lot more than simply buzz around and spread diseases. The Blowflies and flesh flies are a formidable power in nature's recycling scheme. They quickly appear at the scene of a death and can eat even a large Badger carcass in just a week or so.

Hoverflies are the stars of the garden. Appearing to hang in the sunlight by threads, their aerobatics are impressive enough, but consider the pollination services they provide. Their grubs can be found actively scooting about on aphid-infested plants hoovering up 250 or so aphids during their larval career. Many other flies are predators and parasites, all battling away to recycle and remove some real pests from the garden. And flies are also food for things we like a little more, such as bats, Spotted Flycatchers, Swallows and Nightjars.

Robber flies

These are the mean-looking muggers of the fly world, easily recognised by their shady ways and appearance. They have a long spindly body, and long lanky legs equipped with bristles to snare their insect prey.

They shoot off from a perch and catch their prey in mid-flight before returning to the same perch to suck out the contents of their victim.

Cranefly

Walk through long grass or leave a light on in late summer and these gangly insects that seem to be nearly all legs and no body are sure to be seen. The larvae are called leatherjackets and you can often find them in moss.

Winter gnats

A few insects can be seen on the wing even in the middle of winter. These dancing clouds of minute midges are actually a displaying group of males. As soon as a female comes along, they have a little kiss chase before mating.

Left: The fly we all like – the pollen and nectar eating, colourful little chaps we call Hoverflies.

Left: A caddisfly. The larvae and the pupae are entirely aquatic.

Right: Mayfly wings are gorgeous gossamer affairs held 'butterfly like' above the body.

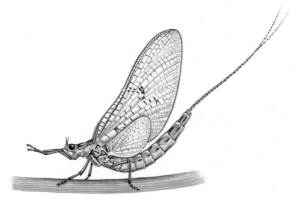

When is a fly not a fly?

Caddisflies, Mayflies and Stoneflies are fly-like flying things, but they are not flies. They are very different insects and not even slightly related to each other.

However, they are together here because they all have nymphs that live below the surface of the water and if you are having a bit of a dabble about you are likely to find them in your pond net.

Caddisfly

These insects are very moth-like in their ways and appearance. Covered in a short, hairy down, this is how they get their family name the Tricoptera, which means 'hairy winged'.

Mayfly

Mayflies don't just fly in May and they are not flies! But there should be no mistaking these winged doilies with other insects hanging around water. Mayflies have three long thread-like tails at the tip of their abdomen and in some species these tails are so long that the adults dancing over the water look as if they are being worked by submerged puppeteers with wires.

Stonefly

Stoneflies can be confused with Caddisflies as adults but one look at their wings which fold flat on the insects' back and you should be able to tell the difference. The scientific name also rather usefully reflects this: Plecoptera means folded wings.

MAYFLY FOR A DAY

On emerging from the water, the Mayfly nymph ruptures to produce an insect that has wings and can use them right away, handy when trout are circling like sharks! This stage, known as a dun, only lasts for a maximum of a few hours before the skin splits again and an even more splendid insect emerges.

The Mayfly is the only insect to have a moult in its winged form. Now it is ready to fly properly and mate. It has no mouthparts and few reserves so the marriage is brief and few live longer than a day. It is this short life span that gives them the scientific name *Ephemeroptera* from the Greek to live a day.

The great thing about water is that you can find insects in it all year around, even in the depths of winter. Stand in the shallows in wellies, upstream from your net. Kick the bottom a few times and you should dislodge quite a few weird creatures to scoop up. Among the assortment of oddities now writhing in your net will be some like gadgets James Bond would be proud of. Many are so small

that they spend their lives in the shelter of objects on the stream bed.

Animals that fall into this category include the tiddly little two-tailed larvae of Stoneflies and the Mayflies (three tails and a rounded body).

Other mayfly nymphs are masters of the fast stuff and can be found in the most rapid riffle, because they can hug the contours of a pebble, not even raising an antenna to risk being whipped from their holdings.

Dip a net through the silty regions of a stream and you will find members of the same groups of insects but they will be more tube-like. Like worms, they burrow in the sediment. Some, like the mayfly Ephemera danica, have gills on their back so they can breathe while they are buried.

The best way to find Caddisfly larvae is to turn the contents of your net out into a dish. If a Caddisfly larva is present, it will usually blow its cover by moving after the water has settled. Most slowly amble from place to place, towing with them an ornate case of stones, snail shells, twigs, reeds or sand, depending on the species and the habitat.

The Scorpion Fly

This is another striking fly-like creature that isn't a fly. It has four large wings, each marked with distinct dark blotches. Its bright yellow body is long with black bands and it has a head with red eyes and a long pointy beak, with its jaws on the tip. To add to its rather scary looks, the male has huge pincer-like claspers on the tip of its abdomen, which are often held up and arched slightly over the insect's back. Look out for it resting on leaves in hedgerows and gardens.

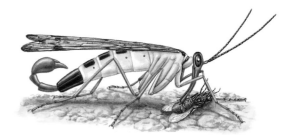

Above: The Scorpion Fly is another fly that isn't. They are named for the appearance of the male's reproductive organs on the end of the abdomen; females have a thinner profile.

Lacewings

In late summer you will almost certainly see one of these green night fairies in a window of your house. They are special insects with large flouncy wings, which they hold tent-like over their backs when at rest. Their eyeballs look like Christmas tree baubles, hence one of their other common names – golden eyes.

Despite their gentle lazy demeanour, these are top predators and another insect that gardeners should welcome. The adults and their larvae are avid fans of aphids. Each one probably consumes a thousand or so in its lifetime.

The larvae can be found among the foliage, disguised as aphid husks. Having sucked the aphid's body dry, they attach the empties to their back – a kind of wolf in sheep's clothing. This beast will eventually pupate in a silken sock hidden in a crack or crevice and emerge as an adult insect to start the aphid onslaught once more.

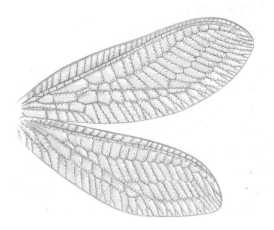

Top right: Lacewing camouflage can be pretty effective!

Middle right: The eggs of Lacewings are quite bizarre. They look like a set of pins balancing on their points.

Right: A Lacewing's chiffon-like green wings.

The little 'uns

This book has mostly been about the 'big game' of the insect world and not a lot of attention has been given to the really small invertebrates. But they are out there in huge numbers.

a pseudoscorpion

Silverfish

You probably wouldn't expect to find a living fossil under the kitchen sink, but here in damp, cool corners of our homes Silverfish survive. Look at a Silverfish and you will be staring at an insect that would have skittered around the feet of dinosaurs in a virtually unchanged form.

These primitive insects could be the ancestors of all insects. Silverfish have survived because they eat pretty much anything organic from plant and animal remains to fungus, mould and even old glue and wallpaper paste.

Silverfish are rarely a pest and tend to go unnoticed unless one gets trapped in a bath or sink during its nocturnal wanderings. The silver body is clothed in a layer of pale scales that brush off easily – hence the common name. The multiple tails identify them as a member of the Thysanura – the bristletails.

Right: Silverfish scamper among the cleaning stuff, dusters and old sponges under the sink.

Pseudoscorpions

Pseudoscorpions are great looking beasts (see page 57). I bet that you have them in your own garden, school playground or park hedge. They are not real scorpions; they have no sting in the tail, they are only 2–5 mm long and they are completely harmless unless you are about 1 mm long! Prey includes small creatures like mites, nematodes and springtails. The pseudoscorpion overpowers them with a lethal injection given through the tips of its hollow pincers.

Springtails

Springtails get just about everywhere and occur in huge numbers. They are very variable in colour and shape. Some are rounded, others are long and thin, some are dark and velvety looking and others are bright vivid greens and yellows. But despite all this potential confusion, if it is small (2–5 mm), has six legs, no tail and jumps when you touch it, you've definitely found a springtail.

Pop a couple into a specimen jar and look at their undersides. You will see the mechanism that allows them to pop off so fast when disturbed. Attached at the tip of the abdomen and held folded against its belly is the spring or furcula, a limb that looks a bit like a two-pronged fork. This is clipped in place until needed. Then it is released, pinging down and backwards. The springtail is catapulted forward and away, leaving any slow predator kissing air!

Go out into the garden and take a handful of soil or leaf litter from under the hedge. You will be holding in your palm a whole world of lots of tiny little creatures. A Tullgren funnel will help you meet them. Be sure to have your pooter, lots of specimen pots and field guides, and polish up a good high-powered magnifying lens.

YOU WILL NEED:

a plastic funnel • a collection vessel (a jam jar or any slippery sided pot will do) • white tissue paper • leaf litter • a bench- or table-light

1: Scrunch up the tissue paper and put it in the bottom of the jar. Pop your funnel in the top of the jar.

2: Place the leaf litter in the funnel and shine the lamp close to it from above (but not so close you risk burning the house down).

Now wait for a few hours. Your Tullgren funnel is recreating two conditions that tiny insects cannot tolerate: heat and light. The animals will move down into the funnel and drop into the collecting jar, saving you hours and hours of sieving through leaves to find them.

Further reading

Albuoy, Vincent, and Chevallier, Jean
Nature by Night
New Holland Publishers, 2008
ISBN 978 1 84773 114 2

*Bloomsbury Concise Butterfly
and Moth Guide*
Bloomsbury, 2014
ISBN 978 1 4729 0996 1

*Bloomsbury Concise
Insect Guide*
Bloomsbury, 2014
ISBN 978 1 4729 0997 8

Chandler, David
All About Bugs
New Holland Publishers, 2008
ISBN 978 1 84773 051 0

Carter, David J.
*Collins Field Guide: Caterpillars of
Britain and Europe*
Collins, 2010
ISBN 978 0 00219 080 0

Gibbons, Bob
*Field Guide to Insects of Britain and
Northern Europe*
The Crowood Press Ld, 1996
ISBN 978 1 85223 937 4

Chinery, Michael
*Collins Complete Guide – British
Insects: A photographic Guide to
Every Common Species*
Collins, 2009
ISBN 978 0 00729 899 0

Forey, Pamela, and
McCormick, Sue
Butterflies (Identification Guides)
Flame Tree Publishing, 2007
978 1 84451 841 8

Keller, Laurent, and Gordon, Elizabeth
The Lives of Ants
OUP Oxford, 2010
ISBN 978 0 19954 187 4

MacLarty, Sally
Colouring Bugs
New Holland Publishers, 2009
ISBN 978 0 84773 525 6

Nancarrow, Loren, and
Hogan Taylor, Janet
The Worm Book
Ten Speed Press, 1998
ISBN 978 0 89815 994 3

*New Holland Concise Garden
Wildlife Guide*
New Holland Publishers, 2011
ISBN 978 1 84773 606 2

Packham, Chris
*Chris Packham's Back Garden
Nature Reserve*
Bloomsbury, 2015
ISBN: 978 1 4729 1602 0

Porter, Jim
*The Colour Identification Guide
to Caterpillars of the British Isles
(Macrolepidoptera)*
Viking, 1997
ISBN 978 0 67087 509 2

Roberts, Michael J.
*Collins Field Guide:
Spiders of Britain and Northern
Europe*
Collins, 2001
ISBN 978 0 00219 981 0

Taylor, Marianne, and Young, Steve
Photographing Garden Wildlife
New Holland Publishers, 2009
ISBN 978 1 84773 486 0

Taylor, Marianne
401 Amazing Animal Facts
New Holland Publishers, 2010
ISBN 978 1 84773 715 1

Teyssier, Jean-Claude
Amazing Insects
New Holland Publishers, 2009
ISBN 978 1 84773 516 4

Tolman, Tom, and Lewington, Richard
Collins Butterfly Guide
Collins, 2008
ISBN 978 0 00724 234 4

Walliser, Jessica
*Good Bug, Bad Bug: Who's Who,
What They Do, and How to Manage
them Organically (All You Need to
Know about the Insects in Your
Garden)*
St Lynn's Press, 2008
ISBN 978 0 97646 319 2

Waring, Paul, Townsend, Martin,
and Lewington, Richard
*Field Guide to the Moths of Great
Britain and Ireland*
British Wildlife Publishing Ltd, 2009
ISBN 978 0 95313 998 0

Wood, Lawson
*Sea Fishes and Invertebrates
of the North Sea and English
Channel*
New Holland Publishers, 2008
ISBN 978 1 84773 125 8

The Amateur Entomologists' Society
PO Box 8774
London SW7 5ZG
aes@theaes.org
www.amentsoc.org

Bees, Wasps and Ants Recording Society
Membership Secretary:
David Baldock
Nightingales
Haslemere Road
Milford
Surrey GU8 5BN
www.bwars.com

British Arachnological Society
Secretary:
Mr John Partridge
31 Duxford Close
Redditch
Worcestershire B97 5BY
Tel: 01527 544952
secretary@britishspiders.org.uk
http://wiki.britishspiders.org.uk

British Butterfly Conservation Society
Manor Yard
East Lulworth
Wareham
Dorset BH20 5QP
Tel: 01929 400209
Fax: 01929 400210
info@butterfly-conservation.org
www.butterfly-conservation.org

British Dragonfly Society
Secretary: Henry Curry
British Dragonfly Society
23 Bowker Way
Whittlesey
Peterborough PE7 1PY
bdsecretary@dragonflysoci.org.uk
www.dragonflysoc.org.uk

Brunel Microscopes (BR) Limited
Unit 2, Vincients Road
Bumpers Farm Industrial Estate
Chippenham
Wiltshire SN14 6NQ
Tel. 01249 462 655
Fax 01249 445 156
mail@brunelmicroscopes.co.uk
www.brunelmicroscopes.co.uk

Buglife (The Invertebrate Conservation Trust)
First Floor
90 Bridge Street
Peterborough PE1 1DY
Tel: 01733 201 210
info@buglife.org.uk
www.buglife.org

Conchological Society of Great Britain and Ireland
Secretary: Rosemary Hill
447B Wokingham Road
Earley
Reading RG6 7EL
secretary@conchsoc.org
www.conchsoc.org

Gordon's Entomological Home Page
www.earthlife.net/insects/

B&S Entomology Net Suppliers
37 Derrycarne Road
Portadown
Co. Armagh BT 62 1PT
Tel: 07767 386751
Fax: 028 3833 6922
enquiries@entomology.org.uk
www.entomology.org.uk

The Microscope Shop
Oxford Road
Sutton Scotney
Winchester
Hants SO21 3JG

Small Life Supplies
Station Buildings
Station Road
Bottesford
Notts NG13 0EB
Tel: 01949 842446
Fax: 01949 843036
emma@small-life.co.uk
www.small-life.co.uk

Watkins and Doncaster the Naturalists
PO Box 5
Cranbrook
Kent TN18 5EZ
(General naturalist supplies – nets, boxes, moth traps, pooters, etc.)
Tel: 0845 8333133
Fax: 01580 754054
sales@watdon.co.uk
www.watdon.com

The Wildlife Trusts
The Kiln
Waterside
Mather Road
Newark
Nottinghamshire NG24 1WT
Tel: 01636 677711
info@wildlifetrusts.org
www.wildlifetrusts.org

Wildlife Watch
(Contact details for Wildlife Watch are the same as for The Wildlife Trusts)
watch@wildlifetrusts.org
www.wildlifewatch.org.uk

Author's acknowledgements

A big thank you goes out to my parents Sandy and Steve and my brother Paul who have tolerated my multi-legged friends, especially the silk moths in the wardrobe and the Tarantulas under the bed, for the best part of 29 years.

I would also like to thank all those at New Holland especially Jo Hemmings for encouraging me to write this book and to Lorna Sharrock and Mike Unwin for not shouting at me and getting too cross when I give them some pathetic excuse for not meeting a deadline!

My neighbours Rob, Julia, Madeleine and Joe for allowing me and photographer Dave Cottridge to invade their pond, to photograph Southern Hawker dragonflies emerging.

Photographic acknowledgements

All photographs by David M. Cottridge, with the exception of the following:
Melissa Edwards: 6, 11, 25, 61, 65, 84, 95, 104, 105, 118
Richard Revels: 7, 16, 19, 21(r), 22, 29, 30, 34, 48, 54(br), 59, 60(b), 66, 69, 71, 73, 89, 91, 92, 97, 101, 102, 109, 113, 117, 121, 123
Jason Smalley: 31

Artwork acknowledgements

All artwork by Wildlife Art Agency, with the exception of the following:
Sheila Hadley: 26, 32(bl), 58, 106, 112(l)
David Daly: 17, 30

(b=bottom; l=left; r=right)